a **LIVING** FREE guide

Backyard Farming
on an Acre
(More or Less)

by Angela England

ALPHA

A member of Penguin Group (USA) Inc.

ALPHA BOOKS

Published by Penguin Group (USA) Inc.

Penguin Group (USA) Inc., 375 Hudson Street, New York, New York 10014, USA • Penguin Group (Canada), 90 Eglinton Avenue East, Suite 700, Toronto, Ontario M4P 2Y3, Canada (a division of Pearson Penguin Canada Inc.) • Penguin Books Ltd., 80 Strand, London WC2R 0RL, England • Penguin Ireland, 25 St. Stephen's Green, Dublin 2, Ireland (a division of Penguin Books Ltd.) • Penguin Group (Australia), 250 Camberwell Road, Camberwell, Victoria 3124, Australia (a division of Pearson Australia Group Pty. Ltd.) • Penguin Books India Pvt. Ltd., 11 Community Centre, Panchsheel Park, New Delhi—110 017, India • Penguin Group (NZ), 67 Apollo Drive, Rosedale, North Shore, Auckland 1311, New Zealand (a division of Pearson New Zealand Ltd.) • Penguin Books (South Africa) (Pty.) Ltd., 24 Sturdee Avenue, Rosebank, Johannesburg 2196, South Africa • Penguin Books Ltd., Registered Offices: 80 Strand, London WC2R 0RL, England

International Standard Book Number: 978-1-61564-214-4
Library of Congress Catalog Card Number: 2012941776

14 13 12   8 7 6 5 4 3 2 1

Interpretation of the printing code: The rightmost number of the first series of numbers is the year of the book's printing; the rightmost number of the second series of numbers is the number of the book's printing. For example, a printing code of 12-1 shows that the first printing occurred in 2012.

Printed in the United States of America

Note: This publication contains the opinions and ideas of its author. It is intended to provide helpful and informative material on the subject matter covered. It is sold with the understanding that the author and publisher are not engaged in rendering professional services in the book. If the reader requires personal assistance or advice, a competent professional should be consulted.

The author and publisher specifically disclaim any responsibility for any liability, loss, or risk, personal or otherwise, which is incurred as a consequence, directly or indirectly, of the use and application of any of the contents of this book.

Trademarks: All terms mentioned in this book that are known to be or are suspected of being trademarks or service marks have been appropriately capitalized. Alpha Books and Penguin Group (USA) Inc. cannot attest to the accuracy of this information. Use of a term in this book should not be regarded as affecting the validity of any trademark or service mark.

Most Alpha books are available at special quantity discounts for bulk purchases for sales promotions, premiums, fundraising, or educational use. Special books, or book excerpts, can also be created to fit specific needs. For details, write: Special Markets, Alpha Books, 375 Hudson Street, New York, NY 10014.

Publisher: *Mike Sanders*	**Copy Editor:** *Tricia Liebig*
Executive Managing Editor: *Billy Fields*	**Cover/Book Designer:** *Rebecca Batchelor*
Senior Acquisitions Editor: *Brook Farling*	**Indexer:** *Celia McCoy*
Development Editor: *Lynn Northrup*	**Layout:** *Brian Massey, Ayanna Lacey*
Senior Production Editor: *Kayla Dugger*	**Proofreader:** *John Etchison*

Cover Images:
Antique handle, vegetable garden in the background © Superstock
Vegetables © Masterfile

USA, New York, Flanders, Hens in hen house © Masterfile
Hand with knife taking out beewax from honeycomb, close-up © Masterfile

Contents

Appendixes

Foreword

Of all the times to be living off the land, I think that today is the most exciting. Savvy farmers who came before us and the modern homesteaders of today have paved the way for today's backyard farmers who have even the most modest amounts of land. Families that are drawn to farming, looking for better nutrition and excellent flavors, find that it's no longer necessary to have acres of land to reach their goals.

We're incorporating small-space techniques such as vertical gardening to grow our vegetables—which allows us to grow in the narrowest places. Experimenting horticulturists have brought us petite-sized fruit trees compared to its full-sized cousins, which offer a delicious bounty in a miniature landscape. There's been a renewed interest in keeping livestock such as dairy goats, chickens, and rabbits.

The backyard farmers of today enjoy all the good parts of growing, raising, and harvesting their own food—along with other necessities—without the fear of a potentially dangerous new land, or the physical hardship that farmers of the past endured. In general, we're more independent due to advanced modern technologies.

It's a powerful thing to realize that we can actually provide the basic necessities for ourselves and our families. Yet rising food prices, control of the chemicals that we consume, plus farm-fresh goodies are only a part of the picture. Self-reliance and handicrafts feed our souls. I believe this sense of empowerment spills over into the rest of the areas of our lives and gives us hope and happiness. Of course, at the very core of it all is the pure joy of being able to offer these things to our family and friends—it's perhaps the sincerest act of love.

Americans are waking up, becoming wiser, healthier, and taking a more hands-on approach to self-reliance. Many are pleasantly surprised to notice that in their search for a more self-reliant lifestyle, they are more like their farming ancestors than they may have thought.

With her commonsense strategies and step-by-step guidance, Angela England brings you a book that will inspire you to plant, raise, build, harvest, make, and preserve home-grown bounty from your own backyard … all in an acre, more or less.

—Chris McLaughlin, author of *The Complete Idiot's Guide to Small-Space Gardening* and *Vertical Vegetable Gardening (A Living Free Guide)*

Introduction

"Daddy, what is it?" she exclaimed, squinting in curiosity and horror at the prickly vine that had swallowed up her play area in the backyard. End-of-school madness had given way to June's warm California sun and she was free to play for the summer, but the intruder had sprouted up seemingly overnight. "I don't know, honey, it looks like some kind of vine. We should let it grow and see what it turns into."

And so they did. Through yellow trumpet-shaped flowers, and the beginnings of a small green fruit, they wondered what would appear on their mystery vine. As the round ball of fruit grew larger, and took on a yellowy-orange hue, they realized that last year's Jack O' Lanterns had left a persistent volunteer in the yard.

That child was me—growing up in my suburban California home meant more time in the pool than the garden. However, the fascination of watching that single, rogue pumpkin grow to full maturity, and take its turn as a Jack O' Lantern later that fall, sparked in me a love of growing things that would not be realized until many years later.

I met my husband at college more than a decade later, and I remember the first time I visited his house realizing how very different our childhoods were. While I grew up playing in the swimming pool, he was feeding cows and fixing fences. While I could walk to the corner store and back in just a few minutes, he lived miles from town and made a monthly trip that was a thrilling change of pace. We were miles apart, but we both had a lot to teach each other.

Once we married and had children we began to seek a lifestyle that was more and more independent and in balance with God's design. Each year we've taken bigger steps down the path of self-sufficiency. Beginning with a small home just inside city limits, with less than a half-acre plot, we've found ways to increase the amount of food we produce for ourselves each year. We started with a simple garden and each year have added more elements to our backyard farm.

Backyard Farming on an Acre (More or Less) is a direct result of that journey and my desire to share what can be accomplished, no matter where you live. Backyard farming is about creating a healthier family, a healthier food system, and a healthier Earth for yourself and those who come after you. It's about recapturing the balance of years gone by, married with the knowledge and breakthroughs of modern ages. Everyone's journey into backyard farming is unique, and that's the beauty of it.

My hope is that you'll be inspired, challenged, and informed through what you read here. Please don't feel that you have to try everything right off the bat, or that you somehow fail as a backyard farmer if you don't have any desire to try one of these elements. Each of these is a choice, an option. Part of the homesteading buffet that you can select from based on your personal situation, desire, time, energy, and family dynamics.

So pick one thing to start, nibble on that for a while, and if it excites your taste buds, go back for seconds! My beginning was a simple rogue pumpkin seed that consumed half my childhood backyard. What will yours be?

Acknowledgments

This book is the work of so many people it hardly seems fair mine is the only name on the cover. I'm grateful to so many, for more than it's possible for this wordsmith to express.

To my parents, for instilling in me a love of learning, and for always indulging my many curiosities and interests, no matter how odd or fleeting they were.

To my husband, Sidney, for teaching me where my food comes from and helping me learn to embrace this new lifestyle. For always being there as the rock who holds me steady through my many moods and brilliant ideas. For being an amazing father to my children, and a man of God who lives his life with integrity and honor.

To my kids, for being the driving force behind my desire to change the world, and for teaching me to look beyond my own selfishness and see generations in the future. For reminding me that I'm only borrowing the world from my grandchildren.

To my friends, who supported me through frantic Facebook and Skype breakdowns, panic attacks, and celebrations. Alli, glitter and brilliance who believed in me and saw the potential before I did, I'm so glad I let you hire my clone. Brannan, who is the best website editor a gal could ask for, I appreciate you keeping Untrained Housewife alive while I sunk into this project. Amanda, who makes me look better than I really am and puts the polish on my websites, thank you for knowing when to pester me again. Heather, my special snowflake, who is always available and will remind me that I am strong, human, and able. Blissfully Domestic Community, the writers and editors whom I love working with, thanks for pulling up the slack when I was distracted and for supporting me these past few weeks. Becky, for taking my rambles and pencil sketches and making gorgeous art. Dorothy, the best VA in the world, thanks for working on any task assigned, big or small, with a cheerful attitude. Chris, my partner in writing crime, I'm grateful to you for putting my name out there in the first place and sharing the delight of book writing with me.

And most of all to the Lord, who shows me more of His mercy and kindness each day, I'm forever grateful He instilled in me a desire to be a good and wise steward of the earth.

Special Thanks to the Technical Reviewer

Backyard Farming on an Acre (More or Less) was reviewed by an expert who double-checked the accuracy of what you'll learn here, to help us ensure that this book gives you everything you need to know about backyard farming. Special thanks are extended to Jodi Hein.

Trademarks

All terms mentioned in this book that are known to be or are suspected of being trademarks or service marks have been appropriately capitalized. Alpha Books and Penguin Group (USA) Inc. cannot attest to the accuracy of this information. Use of a term in this book should not be regarded as affecting the validity of any trademark or service mark.

Living Large on a Small Scale

In this part we look at the importance of backyard farms on the overall health of the economy, health, and stability of our world. You'll learn some stark statistics about the damage being caused by our current food production methods. You'll also learn the many benefits of a small-scale backyard farm.

After you've seen why this is an important step of self-sufficiency, our attention will turn to the all-important factor in any backyard farm: the land. We discuss your options for acquiring your backyard farm: buying something already established, starting from scratch, or adapting what you currently own. You'll learn what to watch out for and look for in a new property.

When you have an idea of what you want to try your hand at, you'll be encouraged to create an in-depth plan, learn how to assess your current situation, and create a plan for achieving your backyard farming goals.

The Backyard Farm Adventure  1

When you grow up watching *Little House on the Prairie* like I did, you develop a rather romantic notion of homesteading and "living off the land." It seems like every day would be an exciting TV-episode-worthy adventure. The truth is a little less glamorous—and a lot more rewarding—than that.

The truth is tending a garden and raising animals is hard work. Rewarding work, no doubt about it. But it's hard. You can't neglect a backyard farm for a week while you go on a luxury cruise. Your livestock doesn't care whether you have the flu or your boss has you working overtime. There are weeds to be pulled, tomatoes to be harvested, and, especially, animals to be fed.

Unlike other authors, I won't try to tell you to avoid this adventure if you are a disorganized person. I am a disorganized person. I am also committed to providing the highest-quality food for my family, on a livable budget. So for many of our food items, producing as much as we can for ourselves is the best way to save money while not sacrificing quality. If you are motivated to take even the smallest step toward an independent and self-sustaining lifestyle, you should! Together we'll explore ideas for how to make the most of your situation.

You might think, *If this is so much work, why bother?* or *Why do I need to produce any of my own food when I can just run to the store and buy some?* There are several reasons to tackle this task, which I'll discuss in-depth here and throughout the book. Please don't feel like you have to jump into everything I'll talk about here. But even one step toward more self-sufficiency is a step in the right direction.

The Growing Food Crisis

There is a growing food crisis in America, and much of the world. As the cost of fossil fuels increases, and the availability of those limited resources decreases, the fact that most of our food is shipped thousands of miles before it reaches us becomes more and more important.

OVER THE GARDEN FENCE

According to the Leopold Center for Sustainable Agriculture at Iowa State University, most food in America is shipped 1,500 to 2,000 miles before reaching the end user—you.

With a single provider sending produce or meat to millions of people, the risks are compounded when something goes wrong. For example, in a highly publicized case in 2011, a cantaloupe recall occurred due to *Listeria* contamination. When a single farm experienced a *Listeria* contamination, the outbreak spread across a minimum of 24 confirmed states. More than 140 people were known to have been infected, although it's possible that not all cases were diagnosed or properly reported.

The large-scale and widespread nature of our food supply makes it more tenuous than most people realize. A war breaking out in a country thousands of miles away, or a drought on the other side of the country, can severely disrupt our commercial food supply. While most Americans went without juicy and delicious melons during the cantaloupe recall, my family was enjoying the last of our cantaloupe and watermelon crop worry free. And while cantaloupe is hardly a must-have food, the potential problem is present in most of our food products in the United States.

From an economic perspective, the United States is now more dependent than ever on foreign imports for our food. According to the U.S. Department of Agriculture, we have experienced an increase of imported fruits, vegetables, and grains of more than 100 percent in the past 20 years alone. And if shipping costs continue to rise, and global drought and weather upheavals continue to decrease food production, we will see the cost—and precariousness—of these imports rise dramatically.

In fact, CNN Money (money.cnn.com) ran a report from World Bank in April 2011 that the price of food had risen more than 30 percent overall in just a single year. A record number of people across the globe fell into the poverty level because of this food pricing increase. For most people, when they are entirely dependent on what's available on the grocery store shelf, these numbers can be frightening.

For others who, like myself, have chosen to produce as much as possible in their own backyard, these numbers are less frightening. I live on a corner lot in a rural Oklahoma town. My front yard looks like an average suburban home, and my generous corner-lot backyard brings the full size of my home to just under a quarter acre. But in that quarter acre we can produce around 40 percent of our food needs and still leave room for the kids to run and play.

What would the difference be if everyone produced even a small garden and kept a backyard flock of chickens for eggs? Well, we ran that experiment once in our nation's history. In the 1940s, spurred by Eleanor Roosevelt's Victory Garden at the White House, more than 20 million American homes had home gardens. These efforts produced as much as 40 percent of the produce consumed in the United States. These numbers reflect my own experiences and show that a return to the simple backyard farm can make a huge impact in the level of food available in the United States today.

Eggs collected fresh every day right out of the backyard. It's only the beginning of what is available to you right outside your back door if you're willing to stretch your comfort zone a little.

An Earth-Friendly Lifestyle

This isn't about a fad or "going green" or anything like that. Backyard farming to support an Earth-friendly lifestyle is about developing a system that is sustainable for the long term. And part of that is understanding how our current food industry is not.

When I married Sidney and learned more about country living, I discovered what a long-term mind-set homesteaders have, even when operating on a small scale. You rotate your crops because that is better for the land three or four years down the road. You don't pollute your land because that can ruin it for decades, and then what will your grandchildren live on? I think in many ways, getting closer to my food at its source, and paying close attention to the land itself, has made me more aware of the damages wrought by modern agricultural methods.

One of the biggest factors in this destruction is the separation of plant production and meat production. We have a system where thousands upon thousands of animals are housed and raised in very small spaces, creating a huge amount of manure and waste. Meanwhile we have farmers growing a huge amount of grain crops on soil that is being rapidly depleted, causing them to use synthetic and petroleum-based fertilizers that are much more costly to the earth and the health of the land. The traditional, historical model was much more simplistic on one hand, and infinitely more complex at the same time.

OVER THE GARDEN FENCE

According to a 2005 study by Cornell University, organic farming producing the same yields of corn and soybeans as conventional farming used 30 percent less energy, less water, and no pesticides. The 22-year trial study concluded that organic farms reduced local and regional groundwater pollution because they did not use agricultural chemicals.

In our little backyard, for example, the soiled bedding from the goat shed is shoveled into a large compost pile. Muck from the chicken coop is also added to the compost bin along with the few plant-based scraps that aren't eaten by either goats or chickens. In the fall the goats are turned loose in the garden area to clean up any weeds or leftover debris from the garden, and then the chickens get their turn to scratch around. They consume grubs and pests, while naturally keeping the soil fertile. The entire bed is covered with any hay the goats have failed to eat so that isn't going to waste, and the garden increases in fertility for the following season. All this happens without a single drop of petroleum-based fertilizer needing to be applied.

This wheelbarrow of straw and manure from out of the chicken and goat pens is on its way to the compost bin. In six months it will feed our soil, which will in turn feed us. If properly managed, the backyard farm can be a mini-ecosystem working in harmony.

From the chickens' perspective, they are given the chance to live in the healthiest way possible, doing what chickens do. Their waste is quickly composted or scattered gradually throughout the property in their moveable pen, and never has the ability to build up to such noxious levels as to become a pollutant and damage the land.

ON A DIFFERENT SCALE

Even if you live on a smaller area than we do and can't let your chickens free-range at all, you can still implement a system for using their waste sustainably. Keep two compost areas going and use the aged compost from one bin in your garden area, raised beds, or container gardens, while you are filling up the second compost bin. Keep plenty of brown matter to mix in with your chicken manure to prevent any odors from building up. (See Chapter 6 for more information about composting.)

The benefits to the earth are even more extreme when you consider the savings in fossil fuels. In their book *Ready, Set, Green; Eight Weeks to Modern Eco-Living*, Graham Hill and Meaghan O'Neill discuss how a small-scale farm utilizing entirely organic methods will use 60 percent less fossil fuel per pound of food than its conventional industrial counterpart.

We've all heard about the growing ecological expense of modern food. Author and food writer Michael Pollan put a heavy number on it when he stated that agricultural systems use 10 calories of fossil fuels for each food calorie produced. In 1940, the trend was reversed, with 2.3 calories of food for every calorie of fossil fuel used. It should be clear that this trend is unsustainable.

As you'll see when we discuss tools of the backyard farm, my husband and I have been successfully growing a large portion of our own food for years now with a minimal ecological cost. If we run the motorized tiller it's only once a year. No tractor. No petroleum-based fertilizers. No toxic waste to be hauled off or contaminate the water supply. No extensive irrigation system to tax the water supply. The smaller scope and varied crops of our backyard farm mean that we are working with a natural cycle instead of against it.

ON A DIFFERENT SCALE

In a perfect world we would see even the major industrial agricultural complexes adopting these organic principles. A 2007 study by the University of Michigan, "Organic Agriculture and the Global Food Supply," concluded that organic farming techniques could as much as triple production in developing nations. Most (but not all) crops are equal in production when organic and conventional methods are compared, according to a 2007 article by the Food and Agriculture Organization of the United Nations titled "Can Organic Farming Feed Us All?" It is totally possible for sustainable and Earth-friendly methods to be implemented on a global scale.

Healthy soil, like what you'll create in your backyard farm, retains water better than poor soil. A thick layer of mulch helps conserve water even more. And in your small-scale farming practices, it's easier to be wise and careful with your watering practices.

Our region experienced a severe drought in summer 2011, and by July we were under watering bans. We were able to use gray water from bathing and cooking, as well as dipping into the small, aboveground swimming pool we could no longer refill, to keep our garden growing well throughout the rest of the season. Compared to larger ranches, like the one my in-laws owned which was without water and lost the majority of that year's crops, our backyard farm produced more than enough for a summer's bounty plus plenty to store over the winter.

Cost and Health Considerations

There are so many considerations to take into account when it comes to the ultimate cost of the food we eat. Entire books and documentaries have been devoted to the topic and I'll not reiterate that here. What I want to share is my experiences with the cost of backyard farming.

I've found that we save a tremendous amount of money on our produce compared to what's available in the local grocery stores. We tend to break even on the overall cost of our meat, mostly because we get the hay for our goats from the ranch my in-laws own. For urban backyard farmers who are importing hay or feed year-round, you might find that your meat prices increase slightly.

There are so many factors that go into what this adventure will cost you—property taxes; the cost of living in your area; the initial investments in things like livestock, fencing, or other equipment; and quite frankly, how "fancy-schmancy" you want to get with it. Do you need a top-of-the-line chicken coop that costs more than your lawn mower? Then chances are you won't save any money on your poultry and eggs compared to the prices you'll see in the grocery store. But I can tell you that last year we put 38 chickens in the freezer at an average of $5\frac{1}{2}$ pounds each for about $.50 per pound altogether.

Here's my secret in keeping the financial cost as reasonable as possible. I'm not a purist; I'm a get-it-doneist. I don't tend toward that naturally, but my husband grew up on a country ranch where they can make almost anything from duct tape and old barbed wire. So when we wanted a portable chicken yard, he made one out of some old lumber left over from another project, old chicken wire from our hatching-out pen two years ago, and a tarp that has seen better days. And you know what? It does the job more than adequately, and the chickens don't care that it cost us $5 in nails instead of $500 brand new.

A chicken coop doesn't have to be fancy, just suitable for the purpose. This portable one moves easily around the yard, provides safety for our foraging hens, and costs only a few dollars.

When I had a baby in April, I purchased all my plants as started seedlings from a local nursery, and the costs of the vegetable garden were twice as much as usual. Compare that to the year I purchased seeds for everything and spent only $90 for seeds that will last two to three years even if I don't save any! All of these things can make a huge difference in what the costs will be for you, and what the savings might be for your family.

A 1996 study by H. Patricia Hynes, an author and professor at Boston University, showed that for every $1 invested in a community gardening area, there is $6 worth of vegetables harvested. I estimate that we save about $400 per month on our grocery bills for our large family because of the food we produce for ourselves, both from the garden and the livestock. Your situation will be unique, of course, but through the various parts of this book I'll share any tips that I've gathered along the way to help you save money on this self-sufficient living journey.

Having said that, there are things that are worth so much to me, I would invest in them no matter what the cost. For example, grass-fed chicken from my own farm, or someone I know locally, is something I would invest in even if the cost were significantly more than the grocery store counterpart. Feeding my family nutrient-rich, cruelty-free, and sustainably grown food is important to me, so there are times when we pay a higher cost either in money, or usually in time, to make that happen. There is a comfort and satisfaction that comes from knowing exactly where my family's food came from.

These chickens are being transported from a chicken house for slaughtering. They were extremely crowded and as I followed behind the truck I could see that many were already dead. I followed this truck for more than 50 miles on the highway—extremely stressful and traumatic miles for the chickens.

One benefit that is harder to quantify is the potential savings in medical costs and health bills for your family. Often, positive eating habits mean fewer trips to the doctor!

A study published by the *British Journal of Nutrition* showed that red meat from grass-finished animals tested with much higher levels of omega-3 polyunsaturated fatty acids than grain-fed feed lot animals (the meat that is typically available in grocery stores). It also showed, in what may be the first human study on the subject, that consumers who ate the grass-finished meat showed higher levels of omega-3 fatty acids (the good kind of fat!) in their blood work. This meat also has higher levels of vitamin A, CLA (the good cholesterol), and vitamin E precursors, according to a 2011 article written in *Mother Earth News*. Precursors are the nutrients in food your body uses to create vitamins, and vitamin E precursors are important for cancer-fighting properties and helping to prevent diabetes.

THORNY MATTERS

As consumers begin to educate themselves about the health benefits of pasture-raised beef over commercial feed lot operations, labels are becoming more confusing. One of the things you might see on a label is the term "grass-fed," which you might think would mean the cow was raised only on sunny grass fields. Unfortunately, the term is often used when the cow is raised on a pasture early in life, and then spends several weeks in a crowded feed lot. When looking for healthier meats that haven't been finished on concentrated grains in a feed lot, look for the term "grass-finished." This means that instead of being sent to a feed lot for the last weeks of life, the cow goes straight from pasture to table.

Pasture-raised chickens produce eggs with higher amounts of iron, vitamin A, vitamin B and folic acid, and omega-3s. They also seem to actually have lower levels of cholesterol in them. And with backyard chickens being some of the easiest livestock to keep, even in a small suburban backyard, these are steps that can really add up when it comes to your family's dietary health. Chapter 12 talks more about raising your own chickens.

The other thing to keep in mind is that when I have something ready to eat, I'm more likely to eat it. That is, I don't usually think to myself, "Gee, I think I should eat a summer squash today." I'm like everyone else—I have a sweet tooth and a half. However, if I go out to the garden in the morning, and we have two ripe yellow squashes, I will harvest them and bring them in. Now I feel obligated to eat the squash, instead of buying a bag of processed munchies and eating those as I might otherwise do. Several studies show that gardeners eat more fruits and vegetables than non-gardeners or the average consumer. Surround yourself with delicious, high-quality food that you've invested your time and effort into and you will gradually change your eating habits by default!

And let's not underestimate the health benefits involved in the act of gardening itself. Many studies continue to prove benefits, like the research by Kansas State University that showed improved hand strength and self-esteem even among elderly participants who gardened. A paper published by Community Food Security Coalition cites several studies showing the various benefits of gardening in reducing risks of obesity, heart disease, and even blood sugar levels in diabetics. Who knows, perhaps gardening could be just what the doctor ordered.

The Many Benefits of Small-Scale Farming

I've found that I'm able to enjoy many of the benefits of small-scale farming right in my own backyard—even in a small town. Being a small-scale farmer has huge advantages for those who only work their farm on a part-time basis. Starting small is also a benefit for those who are testing the waters to see what level of this lifestyle they want to adopt. Our family has been slowly adding new elements to our backyard farm each year to see what steps we are comfortable with.

Cost of Land and Equipment

One of the seemingly obvious benefits of a backyard farm is the lowered cost of land. However, you must take into consideration the fact that land near large urban centers is usually much higher in price than land in rural areas. I'll talk more about the costs of purchasing land versus using the property you already own in Chapters 2 and 3.

Regardless of whether you're buying a new place or using your existing place, the costs will be more reasonable when you have a smaller farm. In my area of Oklahoma, for example, you can still find a 1-acre lot of land for anywhere from $1,000 to $5,000. With a modest family-sized home you can find something for less than $100,000. Compared to a large-scale farming or ranching operation that sells for millions of dollars, you can see how reasonable the prices are to start something like this. For most families you can begin the first steps with what you currently have available to you.

Equipment is another huge cost factor. A garden up to a quarter acre in size can be easily managed by hand without the need for motorized equipment. A larger garden size, or for those who aren't as physically fit to turn a garden by hand, even a good-sized rototiller can be purchased for a few hundred dollars.

A backyard farm has no need for tractors, mechanized irrigation systems, or other expensive farm equipment. Even our outbuildings are modest—an open shed for the goats, a chicken house and portable chicken yard for the chickens, and an 8×10 shed to store equipment is all we have on our property for our backyard farm, and all of these my husband was able to build himself.

OVER THE GARDEN FENCE

The cost of land is a lot less when there are improvements needed before you can get started. If you're willing to trade time and effort to, for example, build fences or clear timber, you can often save on the monetary cost of land.

Ease of Maintenance

One of the biggest benefits of having a backyard farm that is smaller in size is how easy it is to maintain! I can turn the chickens out, feed the goats, and look through the garden for whatever is ready to harvest that day in less than 30 minutes. Fencing in the garden was a two-day, part-time job for one person. Even planting our seedlings, potato starts, and seeds can be done in a relatively short amount of time. And weeding and harvesting chores are easily done by hand.

When we experienced the huge drought mentioned earlier, we were able to do some hands-on things to prevent losing our crop that larger farms wouldn't be able to do. The small size allowed us to do things to keep our garden productive.

When it's time to clean out the chicken house, it doesn't seem like a daunting task. It takes a few minutes, fills a single wheelbarrow with muck that goes immediately to the compost bin, and as a result I don't put it off or have to schedule a huge amount of time to devote to these tasks. Milking our single goat takes 15 minutes in the morning—not hours and hours. The small scale of chores on the backyard farm makes them easier to tackle.

Intensive Production Possibilities

When we plant, we cram our garden to the maximum. Because we regularly apply compost, mulch, and amendments to our garden, we keep the soil fertility at a high level. This means we can space our rows closer together and grow more in a smaller amount of space. Our chores of planting, weeding, and harvesting are all done by hand so the garden rows don't need to be kept far enough apart to accommodate tractor tires. Last season we barely had space to walk between the rows.

The other aspect of intensive production we employ is to tie and trellis many of our vegetables, growing them vertically. I hand tie my little cucumber plants if they don't find the trellis on their own, and weave my tomatoes through the cages and fence. It's no trouble at all when I'm already walking through the garden and weeding or tending it anyway.

This hands-on attention just isn't possible on a large-scale farm. Chapters 4 and 7 will share more secrets for maximizing your garden space.

While there are certainly a huge variety of benefits ranging from economical to health-conscious, for me one of the biggest benefits is the sense of accomplishment that comes from achieving something on this magnitude. Sitting down to a meal that was produced almost entirely in your backyard is an amazing feeling. Seeing your children learn to respect the value of life is eye-opening.

Teaching your children the benefits of healthy eating is easier when your day-to-day activities give you a chance to show hands-on appreciation for hard work and good food.
(Photo courtesy of Tim Sackton)

I've come a long way from my initial forays into self-sufficiency. And there's so much left to learn! Each year I push myself to take another step and investigate a new challenge. The benefits of the lifestyle we are adopting are more than I can enumerate and far outweigh the inconveniences.

Finding Land: What Do You Need? 2

Many backyard farmers are adopting a homesteading mind-set of leaving the city and finding a little bit of land to call their own. The thinking is that they can move to the country and live off the land, enjoying an idyllic country lifestyle. The truth is not quite that simple.

When you're moving with the idea of self-sufficiency, the focus becomes the land. Not the house. Not the granite counters or refurbished hardwood floors. The land and its usability. After all, it's a lot easier to replace the countertops than it is to change the basic composition of your soil. Or make the rainfall average increase an additional 5 inches per year.

In this chapter, I examine the things to watch for if you are planning to move locations with a backyard farming mind-set. Chapter 3 discusses how to adapt where you are currently living to a more self-sufficient system without having to move. It's entirely possible that you start where you are, making the most of what you currently have, and later decide to move to an area with more land. It's precisely what we've done in our family and it worked for us.

Purchasing Undeveloped Land

Having trouble finding the perfect spot for your family? Sometimes it's easier, or more thrilling, to just start from scratch. Purchasing land and then building gives you the ability to have a more Earth-friendly home. It also means you can make it the way you want it instead of adapting what someone else has put together.

My husband Sidney and I considered doing this. His father has 40 acres of completely undeveloped land that we considered building on. I call it Wilderness Hill. It's more than a half mile from the nearest road and would have been starting from scratch at the most extreme example. I will share more of what we learned in this chapter so you'll know what to watch out for.

Location and Cost

The average cost of an acre of farmland in the United States in 2011 was $2,400. But that's the average and most areas have a range. Having glanced through a real estate magazine recently, I noticed two different 1-acre lots for sale. One was for sale for $8,000. One was listed at $900. Why the difference? Location!

The $8,000 lot was within city limits of one of the nearby small towns. The $900 lot was way out in the middle of nowhere with no electricity, no real road, no clearing, no water, no fences. The clue came in the description of a "remote and secluded woodland retreat." Whereas the other was listed as "ready to build." And if you think there isn't much of a difference, think again.

We looked at Wilderness Hill—a gorgeous 40 acres of completely private land—and started making a list of what needed to be done before we could even build:

1. Survey the land. The county wouldn't open up the county line until we'd surveyed the land. Cost? $1,500.

2. Have the county open up the county line road. This isn't anything fancy and nothing more than a dirt road along the county line. The fencing and driveways would be our responsibility. Cost? At least $2,000 in top-fill and gravel.

3. Run in electricity. Our electric company would only go the first 1,000 feet without charging us. After that it was $1,200 per post (every 100 feet). Total cost to the projected homesite? $12,000. (Note that solar power systems for an average family home are available for less than $20,000.)

4. Fence in the entire property. Two of the four sides were totally unfenced and the third side only half fenced. Cost? More than $1,500 in wire, posts, and staples.

5. Prepare the homesite. Backhoe, bulldozer, etc. If you're less handy than my husband, you'll need to hire someone to clear the area as well. Cost? $250 to $1,000 per day.

At this point we thought it might be a lot easier to find a piece of land already settled and ready to go. We hadn't even started on the cost of building the actual house yet. Sure it would cost us more at the beginning to purchase already developed land, but would perhaps save us quite a bit of money as well as time when it is all said and done.

Buying land that is already positioned relatively close to electricity, city or rural water, and other amenities will be less trouble but will also be a lot more money. As much as ten times the price. Sometimes this cost difference will be prohibitive and you'll want to begin examining what it's worth to you to become more self-sufficient.

Living in a more rural location will bring a greater diversity in wildlife. It will usually mean much more affordable land and housing prices, and often a greater cooperative atmosphere with your neighbors. Instead of being the only one in the area with a backyard flock, you'll have more experienced poultry keepers around you who can help get you started or give you a pointer or two. On the other hand, being in a more urban location usually means access to a wider choice of schools, jobs, stores, and other amenities. You'll want to accurately assess your family's needs before uprooting them to an area that doesn't have what you need. Relocating your family is a big decision that shouldn't be taken lightly.

OVER THE GARDEN FENCE

It's a huge misnomer that you have to be out in the country somewhere to backyard farm. Nothing could be further from the truth! In fact, before you move out into a rural area, I highly recommend adopting a more self-sufficient lifestyle and backyard farming as much as possible without moving ten states away. Try working a garden in the backyard you have right now and see if you can hang with it for a year. You might detest it and decide to buy all your produce and meat from local, happy farmers. Better to discover that now instead of two years and $200,000 later!

History of the Land

One thing that you would want to be sure to find out prior to purchasing land is the history of the land. As you'll learn later, certain farming techniques strip the land or leave pesticide residues that could hurt the crops you're trying to grow organically. Maybe the area has been mined. And of course, you'll want to find out about the historical climate of the property as we'll discuss in a couple more pages.

The land around our county was primarily logged, which means heavy equipment. Sometimes that can lead to extreme soil compaction that makes it more difficult for tender roots to get into the soil. Soil needs space for air, water, and beneficial microbes to get down deep and bring up the nutrients, and compaction means that all these nice spaces have been squished out of the soil. Extreme compaction can be corrected, but it takes time and effort and means a lot more than simple tilling. You can see how important soil quality is in making a purchasing decision. (Chapter 6 has a lot of information about building soil fertility.)

In other areas you may find that land has been mined or farmed in such a way that much of the rich top soil has been lost. Acre upon acre of farmland has been ripped open and left bare to the ravages of the wind. Protecting this precious soil is part of why we plant cover crops and mulch, as discussed in Chapter 7. Unfortunately, larger commercial operations aren't always so careful.

Pesticide use in the past three years is enough to prohibit certification as an organic production. Not only that, but the residue of those pesticides can remain in the well water and soil for quite some time, especially where you are growing trees, berries, and long-lived perennials.

If your real estate agent doesn't know the history of the land, try talking to neighbors who live nearby. In most rural areas you'll get a detailed history and learn not only who lived there, but how many kids they had, what their dog's name was, and a funny story about that time their mean ole rooster chased the cat up the tree. If nothing else, you're sure to be entertained.

THORNY MATTERS

Don't forget to inquire about unexpected costs associated with the property. Find out what the property taxes are currently, and what they might be in the future after you've built your home and a couple outbuildings. I know that more than one backyard farmer has been shocked when property taxes triple (or more) the year after they build and move in. Don't get caught unaware!

Climate and Growing Season

Another area where bending the ear of your potential neighbors will be crucial is in understanding the climate foibles where you live. Sidney's grandmother used to say, "Never plant your tomatoes before Easter," which gives a strong clue as to the average last frost date on the old family homestead. Because we live just a few miles down the road, we tend to follow her folk wisdom.

Find out about potential microclimates in your prospective neighborhood. Your growing zone helps you determine which plants will survive in your area and are very general numbers, while microclimates are specific climate pockets that vary from area to area even within a single homestead. Are you looking in a low-lying area? Expect longer-lasting frosts and cold pockets. Are you looking on the south side of a hill? Expect longer hours of sunshine and increase your zone by another half. Is there a large lake near you? That will change the climate in your neighborhood in a way that will be different from your regional zone set by the U.S. Department of Agriculture (USDA) for your entire area (see Chapter 6 for more about zones).

As a backyard farmer, pay close attention to things such as water and land quality. While it might be enough for most people to just have a pretty view, that won't work at all for you! If your land has a good well, or small pond or creek, you'll be much better off in the long term.

Different areas (even different states) will give you better growing seasons than others. While intensive and hands-on techniques will allow you to baby your desired plants a little bit, it makes sense to plant what grows best in your area. For example, I will never have a maple syrup stand and create homemade, delicious maple syrup in my warmer region. But pecans grow very well here. So if I absolutely had my heart set on maple syrup, I should not live here in Oklahoma!

Be sure you find out the length of the growing season. If you have your heart set on specializing in heirloom watermelons, for example, but you move to an area with only a 90-day growing season, you'll be out of luck.

Of course you can cheat some climate issues by using greenhouses and cold frames. These things can all extend the growing season and make it possible to bring items from other growing zones. But sometimes it's not worth the extra hassle when you can grow other items so much easier.

Buying Developed Land with the House

If you decide not to build your own home, you can buy a homestead ready to go. Just recently we toured two different home-with-acreage listings. One had a gorgeous home and wooded land that we would need to clear off as we planted. The other had the perfect land setup, but a home that our family has already outgrown. What to do, what to do. There's a lot to consider when you're looking at a move to full-time backyard farming.

Layout and Quality of the Land

Consider the layout of the property you're examining. How is the land arranged? Is it wooded or cleared? If you have to spend the entire first year clearing large areas of the land so you can begin planting, it might be cheaper to look elsewhere. On the other hand, this wooded land is usually a lot less expensive to purchase at the start.

Where is the house located? Remember that in the northern hemisphere, the southern exposures are highly valued. The sunny south side of the home is where the money is in terms of providing you sunlight hours of growing time for your crops. These are all things you should make note of when you're trying to decide on a property.

Soil composition and land layout are other important considerations. Check the quality of the soil from around various areas of the yard. Check the side yards, the backyard areas, and the front yard. Are there a lot of rocks? That can make it difficult to dig and plant fence posts. It can also be a lot of trouble for tillers and growing root crops like carrots and potatoes.

Does the soil have fine particles and heavy clay? This soil will hold water and have poorer drainage. It will take plenty of amendments and added organic material to make it *friable*. Will this be more trouble than it's worth?

DEFINITION

Friable soil is loose, crumbly soil. The term usually refers to fertile soil full of organic material.

Is there a low spot that stays wet and mushy year-round? We have one corner in the back of our property that stays wet and is unusable when there's a strong rain. But it's a low and shady corner that wouldn't grow much anyway, and helps drain out the rest of the yard. If you are looking at a home with a septic tank, drainage and the lay of the land will play an even more important role.

For gardening, a relatively flat area with good drainage, fertile soil, and good exposure (sunshine) is ideal. You'll not always be able to find a reasonably priced property with *all* the elements you need, so that's where you'll want to consider the costs and hassles of fixing up the property for yourself. Would it be easier to add drainage, or easier to create terraces on a highly sloped property?

If you are purchasing a piece of land with the intent of using it to produce food, I recommend that you test the land before purchase. Take soil and water samples for testing (most county extension offices will do tests for a small fee) so you can better judge the suitability of the purchase.

Outbuildings and Existing Amenities

When you're looking over potential properties, you will see a wide variety of outbuildings, sheds, and barns. You might see a house sitting on an otherwise empty lot with no fencing, no shed, no storm cellar, no nothing.

We've seen generous-sized properties with huge mansions that were clearly never meant to house a backyard farm as the house took up almost all the property, and the giant garage took up the rest. The huge amount of concrete and utter uselessness of the setup makes it clear that the property was never intended for housing a garden, chickens, or anything productive. These would be the properties to avoid at all costs because the price of acquiring the bloated properties would be prohibitive and most of the so-called amenities would be rather useless to a family interested in self-sufficient living.

On the other hand, you might find a true diamond in the rough for a very reasonable price. We recently looked at a property that was priced about 20 percent less than others in the area. Why? The house was cluttered, and the property was completely overgrown with a good two or three years' worth of overgrowth. It *looked* horrible at first glance. But the fencing was sound for the most part, the pond was in good shape, there were mature fruit trees already on the property, and a couple outbuildings in excellent, if cluttered, condition. Most people couldn't see past the surface clutter and briars that had grown up. If you can snag something like this for a reasonable price, you'd be much closer to self-sufficient living without a heavy debt hanging over your head.

Take note of any outbuildings in particular. These can raise the asking price of a property a good portion. They also raise the property taxes, so you'd want to ask what taxes tend to run in the county. Sometimes you can go just a mile or two down the road and cross the county line, or out of city limits, and cut your property taxes to a mere fraction of what they would have been.

If the property doesn't have any outbuildings, but has some nice level areas near the house, it wouldn't be hard to add some buildings as needed. After all, if your backyard farming plan only includes a garden and small chicken flock, you wouldn't need a property that included a goat shed, a milking barn, and a couple stalls for sheep or other small livestock. Some of these types of buildings could be easily converted with a few modifications inside, but other times the buildings are built so specifically, you'd spend way too much money in the purchasing of the building for something you will never use.

When being charged in the initial purchase for an outbuilding, shed, or barn of some kind, you really want to look at the quality. Is the roof sound? Is there proper ventilation and is that ventilation screened? What kind of siding has been used? Untreated wood will quickly weather and become exposed to the elements. It would need to be treated with water sealant to be able to be cleaned or withstand a rainy season. Tin siding, on the other hand, offers water resistance, but little insulation when used alone. If it's a thin-sided shed you're hoping to convert to a chicken coop, you might want to plan on adding another layer for insulation inside.

Fencing

One of the most commonly overlooked benefits to a property is the fencing. Your property will need sturdy fencing to protect any livestock from predators. Simple little picket fencing looks nice, but is often pricy to install and doesn't have much practical use. If you are paying the price for what the former owners have installed, only to have to re-fence the entire property yourself, you haven't saved a nickel.

This little picket fence gate looks pretty, but it didn't keep in the dogs, the chickens, or the goats. It now serves as a simple gate in and out of the garden, but has been reinforced with wire at the top.

Wood panel fencing can be difficult to maintain. But wood is often less expensive than some of the other options, such as vinyl. Treating wood will increase the cost, as will painting the fence. If you live near a wooded area, you may be able to procure the resources for free or very inexpensively.

This round pen was made from homemade bodark fence posts. The dense wood has a natural resistance to rotting and allows for a relatively maintenance-free fence. In this case it didn't cost money—only labor.

Vinyl fencing is much more expensive, but after it's set up it doesn't need any maintenance and will last a long time. Vinyl fencing is not the most Earth-friendly fencing material, but it is highly attractive. If the property has vinyl fencing already in place, you'll likely be paying extra for it.

This vinyl privacy fence helps block our garden mess from the rest of the neighborhood.

Regardless of the type of fencing, you'll want to consider the layout of the fencing. Is there already a clear front yard and backyard area? Or is the entire lot wide open?

Are there any cross fences in place? When you house livestock in your backyard farm, you will benefit from being able to move them from one pen area to another. Depending on your region, a good fence can help deter predators, or garden marauders such as deer and rabbits.

Other amenities to consider are things that will increase your self-sufficiency—items you might consider adding to a home later yourself anyway. A woodstove, for example, can be a big asset. We recently looked at a property where a wood-burning fireplace was in the largest room in the house and the blower had been modified to tie directly into the central heat unit. So the central heat only kicked on when the fire died down, saving hundreds in utility costs each winter.

Even without being tied into the central heat's thermostat and duct work, a wood-burning fireplace or stove can give you an alternative source of heat that shaves money off your bills, depends on easily renewable resources, and isn't going to go down during a storm or icy weather. The more rural your property (read that: the further from help when the power goes out) the more important it is to have a backup in case of power outages.

Of course, a wood-burning stove only saves you money if you have access to wood. If there isn't enough timber on your property to burn, you can often trade labor to neighbors. We don't have extra trees on our property, but we can get them for the taking when someone has a tree blown over just by offering to clean it up for them.

Access to Water

Water sources are important to a backyard farmer. A working well is a boon for a homestead to lessen your reliance on city water. Even if the well water isn't plumbed to the house, having an extra source of water for animals and garden is a definite asset. We learned this firsthand when our city put everyone under a watering ban during summer 2011. The only people who still had productive gardens were those who had alternative sources of water.

If well water is the home's only source of water, you'll want to have it tested to make sure it's drinkable. Find out the depth and age of the well; older, more shallow wells may end up needing repair, or more expensively, replacement. Also find out the well's flow rate. If the well isn't tapped into a good water source, you could drain it of water reserves during regular daily use faster than it is able to be refilled. In this case, you would need water tanks to hold rainwater for things like irrigation and livestock to prevent overtaxing your well.

Rural water systems should also be tested as the water can have contaminates or chemical levels that aren't healthy. Our local rural water often tests high in nitrates, which can have negative effects on livestock (and people!), especially pregnant females and babies.

If the land you're looking at has a creek or small pond, watch out for clever language in the description before you assure yourself that it will be a year-round source of water for your animals. Phrases like "flood-zone creek" or "seasonal creek" simply mean, when it rains you have a creek and when it doesn't rain you don't have a creek. However, a creek or pond could be something to add to your list of cons if it takes up a large amount of your useable land. Ultimately, that's what it comes down to when considering a property for a backyard farm. How much of the land is useable? How much can be productive?

THORNY MATTERS

Having grown up in a big city, I never understood the importance of water. I turned the knob on the hose, and water came out. Well, during the summer of 2011, I saw firsthand what can happen when wells fail and water is scarce. The well at my in-laws' house, a 50-foot well that had been hand-dug by Sidney's grandfather 75 years prior, went dry. All of the water for the entire house (cooking, flushing the toilet, bathing) had to be trucked in from a neighbor's house. Know where the water for your property will come from, and be sure the quality *and* quantity is sufficient.

Zoning Restrictions

Of course, finding the perfect place for your backyard farm doesn't do you any good if it's illegal to grow a garden there. It may seem preposterous that someone could actually get in trouble for feeding their own family, but in some areas of the United States, that is exactly what is happening.

Take the much-publicized case of the woman who grew a garden in the front yard and faced misdemeanor charges as a result. Her city planners vowed to pursue the case all the way to court, if necessary, which would have landed her more than 90 days in jail! Thankfully the city dismissed the charges when internet outrage was unleashed.

But that isn't the only case in recent history. The resurgence of self-sufficiency has butted heads with archaic or vague restrictions in many suburban areas. Between zoning restrictions, homeowners associations, and cranky neighbors, it seems almost impossible to live on your own land sometimes.

I have an acquaintance who moved into the perfect house with the perfect yard, only to discover that the homeowners association would not allow her to keep chickens, not even behind a privacy fence in her own backyard. She may or may not have smuggled in "voiceless" ducks that don't make as much noise so she could still enjoy fresh backyard eggs. They might live in a tidy little pen behind her shed, privacy fence, and beautiful garden, and she might be at risk of being fined every day as a result. But you'll never hear me give away her secret!

THORNY MATTERS

The only thing more frustrating than wanting to start a backyard chicken flock and being told you can't would be to anticipate a quiet retirement neighborhood and waking up to crowing roosters from your neighbor's unexpected flock. Double-checking your zoning restrictions before making a purchase can keep everybody happy.

There is hope, of course. Many cities and municipalities allow gardens, chickens, rabbits, and other small livestock. In fact, as of December 2011, 93 of the largest 100 cities (by population) in the United States allowed keeping backyard chickens in some form or fashion. Often, cities will limit backyard flocks to hens only, or enforce a smaller size such as six hens per family within city limits. But even with limitations, the fact that so many of the nation's largest cities are friendly toward those seeking additional self-sufficiency is an encouraging trend.

Many cities with currently unfriendly regulations are finding concerned citizens taking up the cause. As regulations loosen and taxpayers make their wishes heard, you will see more and more backyard farms and urban flocks being allowed. But until then, don't make the mistake of putting yourself on the wrong side of the law because you failed to ask questions about zoning restrictions and city ordinances prior to a major purchase.

Building on Land You Already Own 3

You don't have to make a huge life change in order to begin a backyard farm. For most of the people in the United States, it's possible to begin living more self-sufficiently right where you are. Don't wait for "the perfect setup" to get started. Start growing your own food now, even if it's just 10 percent to start. Start doing for yourself. Start lessening your footprint. Start making things for yourself instead of having "buy it from the store" as your default mind-set.

There are a lot of things you can do to begin a backyard farm right where you are. As you read through the other chapters, look for sidebars called "On a Different Scale" with tips for scaling things down to smaller or larger than average spaces. Of course, if you live on more land and are just getting started, you'll want to think about ways to potentially scale things up a notch.

Many urban backyard homesteaders are finding it easy to produce enough of a particular product to not only provide for their own families, but to have enough for market as well. Take some time to browse your local farmers' markets and see what the going prices are for extra produce. You can often get a booth inexpensively and sell your own fruits and vegetables to the general public to help defray some of the costs.

Assessing Your Current Situation

Although your current situation may not seem ideal, most homes you wouldn't consider ideal can certainly be serviceable. You just have to take stock of what you have to work with and think creatively about how it can be adapted to increase your self-sufficiency. There are a lot of very creative possibilities when you allow yourself to think outside the box.

First, assess how much space you have. Include concrete patios and porches in this assessment. Any area that will get around six hours of sunlight or more can be adapted to a productive place. Balconies, window sills, patios, front yards, side borders, and more can be adapted and used to grow fruits and vegetables. Even relatively small backyards can house rabbits, chickens, or bees.

This city home has a thriving apiary with a large beehive and generous honey production.
(Photo courtesy of Nicolás Boullosa)

Let's look at some examples you might see of a backyard farming property.

Quarter-Acre Lot

This first illustration is quite similar to the layout of our own home on a quarter acre. Note that our southern exposure is the entire backyard, giving us plenty of area for growing our garden.

1. Residence
2. Edible landscaped flower beds/herb garden
3. Fruit trees
4. Compost bins
5. Berry bushes
6. Side yard
7. Goat pen

8. Vegetable garden
9. Pecan trees
10. Chicken coop (moveable)
11. Beehives
12. Strawberry pyramid (or perennial vegetable)
13. Garden shed (enclosed shed building)
14. Goat shelter (shares a back wall with the garden shed)

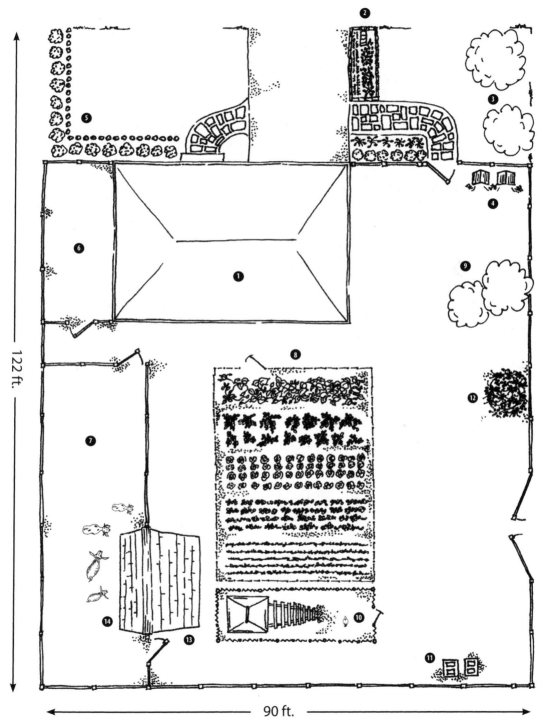

Possible layout for a quarter-acre lot.
(Illustration by Becky Bayne)

We're not utilizing all our backyard space, as you can see from the illustration. With five very active children, we left room for a soccer goal, trampoline, and wading pool. Our garden area is completely fenced in to prevent trampling from the dogs and demolishing from our goats. The goat pen takes up one back corner and the extra side yard gives us a run area for either dogs or goats as needed.

The moveable chicken coop can be placed around the yard as needed. Chickens keep the yard under control. When the fence line grows up, we pen the dogs on the side yard and turn the goats loose. The garden stays active year-round supplied with compost from our bins.

If you don't have kids with a big trampoline and soccer goals taking up yard space, you can increase the garden size. My dream would be to section off the front right corner of the backyard for perennial edibles like asparagus, and berries along the fence line. Raspberries, blackberries, and grapes would be perfect along the fence there! If we had a park nearby where the kids could run and play, I'd probably be able to do that. What resources are available in your local area will help determine what works best for your family.

Even with only a quarter-acre to work with, we manage to produce a wide variety of crops. Our backyard flock of chickens provides eggs and some meat. The two milk goats provide the majority of our dairy for cooking, baking, and drinking (the children especially love goat's milk). Two full-grown dairy goats will also provide two to four kids for meat each year.

With only a quarter-acre we provide hay for the goats on a year-round basis. They eat very little hay during the peak of the spring and summer growing season but eat a good portion each day during the winter. I'd rather pay to bring in outside hay than pay to bring in outside milk and meat.

Half-Acre Lot

This illustration of a sample half-acre lot gives you an idea of what could be grown or raised in this type of space.

1. Residence	6. Chicken coop (moveable)
2. Fruit trees	7. Goat pen
3. Beehives	8. Food crop (e.g., corn, wheat, barley)
4. Herb garden	9. Pecan or walnut tree
5. Vegetable garden beds	

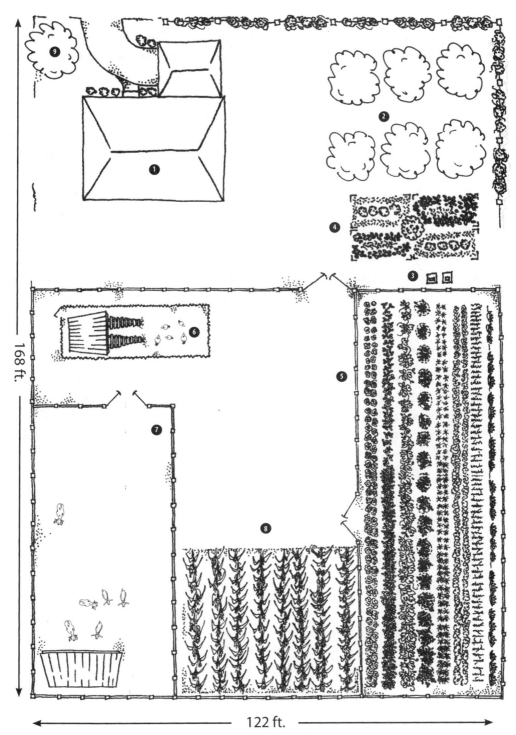

Possible layout for a half-acre lot.
(Illustration by Becky Bayne)

See how the front yard has been converted into an attractive, edible landscape with fruit and herb plants together? Set attractively, you can increase your curb appeal while increasing your productivity at the same time. Figuring out how to maximize your growing spaces in the nooks and crannies allows you the greatest flexibility.

In this case, the backyard is fenced off to allow free-range of the chickens without the flock making a mess in the front yard area or escaping onto the road. The garden area has the ability to be fenced off if needed, giving extra browsing space to the goats. Much less hay would need to be brought into the backyard farm area if the goats were rotated through the extra grassy areas.

And of course, the extra garden plot in the back allows you to grow additional feed. In the illustration it shows corn, but you could grow feed beets, potatoes, or other staples or grains as desired. Wheat could be grown but would produce a smaller harvest compared to more intensively producing feed crops like cowpeas or beans.

There's also room in this design for a greenhouse or cold frames, if desired. These items aren't necessary, but as you'll see in Chapter 7, they will help you extend the growing season. And a longer growing season means more food.

Full-Acre Lot

With a full acre in use, you are able to house multiple animals. Rotating pens will keep most of the animals in grass for the majority of the year, necessitating a smaller hay purchase. The additional space also allows you to grow some of your own staples and feed crops such as potatoes and beets. While it wouldn't be possible to be completely self-sufficient in terms of feed for a large number of livestock, you could get a lot closer with an acre or more in use.

1. Residence
2. Herb garden and fruit trees
3. Goat pen
4. Chicken coop
5. Nut tree
6. Sheep pen
7. Vegetable garden
8. Fruit trees
9. Berry vines (e.g., grapes, raspberries)
10. Greenhouse
11. Strawberry pyramid
12. Garden shed
13. Asparagus bed
14. Beans
15. Potatoes
16. Corn
17. Beehives

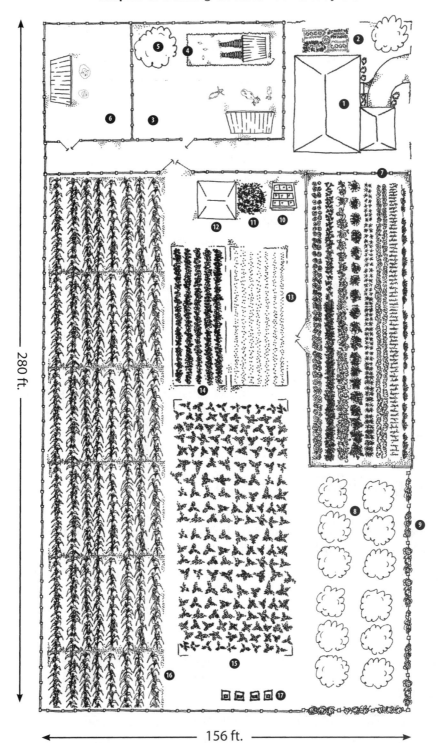

280 ft.

156 ft.

Possible layout for a full-acre lot.
(Illustration by Becky Bayne)

Replacing the back lawn with drought-tolerant food crops will decrease your watering bills. These crops will also make it possible to grow the feed for your animals as well. Notice a large plot for corn in the back of the growing area. There are also large spaces for beans or peas and beets or potatoes.

Alternatively, you could use this area for market-produce. Perhaps you could use this space as a larger orchard for selling jams and jellies, cut-flowers area, or specialty heirloom varieties specifically for higher prices at farmer's markets. The less you keep for yourself and animals, the more you have to potentially buy for your backyard farm, but the trade-off might be worthwhile for you.

Closer to the house are perennial beds for more permanent plantings. Notice the strawberry pyramid, which allows vertical growth of the plants in a set location. You can also enjoy a good size asparagus bed that would be more of a permanent planting as well.

In this example there's a greenhouse, a garden shed for housing tools, a chicken coop, and goat and sheep pens. This is a large number of outbuildings, of course, and not all are necessary for getting started. We don't have this many right now, but if I were creating a long-term plan of what I wanted to have over the next five years, this is pretty close to what it would look like.

OVER THE GARDEN FENCE

The best part about starting slow and adding new things as you're able is that you can test the waters and see what you feel most comfortable with. Do you hate goats? Don't keep them just because we do! Increase the size of your chicken flock instead, or specialize in small sheep grown specifically for their fiber. Whatever it is that interests you and catches your fancy is worth looking into.

Creating a Workable Plan

Consider these illustrations as wish-lists for "someday." Create a few sketches, perhaps in your garden journal, of what you think you'd like to have available over the next year or two. When you have a vision of what you're looking for, you'll be able to move forward with confidence.

This is an acronym I use when coaching blog clients, but it applies to planning anything: you need to have a P.L.A.N.: Precise, Lasting, Arrival, Natural.

Precise

Make your plans precise! Don't be afraid to start planning the number of crops you want to plant. Then figure out how much room they will take up and how much space you'll give them. Now you know precisely how big to build that new garden plot.

Will you have gardens, livestock, or a mix of both? What crops will you want to grow? Part 2 of this book will detail some of the most popular and useful vegetables, herbs, and fruits to grow.

What animals can you house in the backyard? See Part 3 for some ideas of animals you could potentially keep on a smaller scale. Write down your wish list of what to grow and raise so you have a precise picture to work from and aren't wasting your efforts.

Be sure to build a chicken coop the size you need. If you need six hens to supply your family's egg needs and you can only house three hens in your coop, you've wasted your lumber, your time, your money, and your energy.

Wastefulness is almost the opposite of precision when it comes to planning. Often on the ranch we joke that nothing goes to waste. But it's really true—everything serves two purposes, or the leftovers can be repurposed in some way. Not even our time is wasted, as leisurely summer nights spent watching a family movie are spent with purple-hull beans in our hands so we can shell out the beans while we watch the show.

Lasting

The best plan takes place over time and accounts for the effort it takes to achieve a particular goal. It's really easy to say "I want to have a fully self-sufficient homestead that completely supports my family with mature fruit trees, a well-established laying flock of chickens, and milk goats that are ready for milking." That isn't really a plan, though, that's a wish. And because there's no timeline attached to that wish, it's going to feel more overwhelming to actually set out and do.

A plan looks more like this: "This January I'm going to buy my baby chicks so they will be ready to lay in July or August. I'm going to buy all my seeds for the year now while this nursery has free shipping, so I need to find my list of all the crops I want this year."

Maybe this summer you clear out an old shade tree so in the fall you can replant the space with a bareroot fruit tree that will eventually take its place in the landscape, providing your family with food at the same time. Maybe you make arrangements to take your doe goat to a neighbor's house to freshen (breed) so you'll have a baby, and milk, in late winter.

Time is the taskmaster here. Often a plan is the realization that you aren't going to have everything at once. You have to decide which elements need to be done now and which tasks could be done later. You also have to understand when certain things take place on the homestead. What happens each week? What happens each month? Chapter 16 will give you an overview of the seasonal rhythms of a backyard farm.

ON A DIFFERENT SCALE

Your plan doesn't have to be elaborate—especially if you're using your "extra" or fun money to get started, or have an outside income that is supporting your family and paying the bills. If you are taking on the challenge of a backyard farm from a hobby perspective, just create a list, decide on your first big step, and go after it!

Arrival

Arrival means you are heading in a particular direction. These are your specific goals. The sketch you put together of how you're going to adapt your home into a working backyard farm should be clear! Of course, you will never reach that dream ideal in anything closely resembling a straight line. But with a clear-cut vision of what you want to achieve for your family, it will be easier to meet your goals even when unexpected setbacks occur.

To me, this variability of life is part of what makes having a plan so valuable. When unexpected things crop up, we can change the plan a little to accommodate the situation and then move easily to get back on track as soon as possible. For example, I was hoping to buy a greenhouse for the backyard because the plan was to increase the amount of food we were growing. Unfortunately, we had a situation come up with my husband's work truck and it needed to be repaired. Bye-bye greenhouse money. Part of the repairs was a new set of tires on the truck, and I used the old tires to grow potatoes in a tower. It was amazing how many we produced in that very small area! Not to mention the fact that growing the potatoes vertically freed up two rows in the garden for more tomatoes.

Natural

This is the aspect of planning that I want you to keep firmly in mind as you read this book. You will be most successful when you work toward what is a natural fit for you and your family. Your plan should reflect *you*. All the ideas and suggestions are things you can do—things you might *want* to do, not things you *have* to do. So select a goal that makes sense to you. Start small if you have to. Go all out if you want to. Make your plan work for you, and you'll be surprised how far you'll come when you look back a couple years from now.

Write out your plan and start working out the most important elements of backyard farming to add first. For many people, that is a garden and a small flock of chickens. And that's a great place to start. Then note what you need for those elements to be successful. Assess what you currently have and go from there.

ON A DIFFERENT SCALE

If you're taking on a backyard farm as a business and planning to support your family without any additional employment, you need to be a lot more detailed in your planning. Some market research, pricing information within your region, and estimated costs versus income projections will be needed.

Your garden will need a flat, well-drained, sunny area. If you don't have one, you'll have to create one. This might mean removing an out-of-place shade tree, digging up some lawn, building raised beds over a portion of the cement patio, adding containers to your balconies and window sills, or even just making reservations for a space at your local community garden.

Chickens will need housing of some kind. They will need space to roam and forage. They will need protection from predators. All of these elements may have to be added to your backyard if they aren't already present. (See Chapter 12 for information about keeping chickens on a small scale.)

Adding Buildings and Fences

The reason I encouraged you to create your goals and your plan before I talked about changing the *hardscape* on your property is because these structures are semi-permanent. Sure, you can move a fence, but as my husband can tell you from many years of experience, it isn't an easy task. Nor is it exactly fun. Better to plan carefully, do the hard work once, and enjoy it for many years to come. (See Chapter 2 for more on fencing options.) Add buildings carefully and consider the site before you begin building. There are a few things you'll want to pay attention to when you build a new outbuilding.

DEFINITION

The **hardscape** refers to nonplanted elements of the landscape such as sidewalks, buildings, arbors, gates, and fences.

Drainage

You want to make sure that an outbuilding won't be in a low spot where water will drain into the shed and keep everything wet or keep your chicken coop wet and make your chickens sick. Concrete flooring in an outbuilding is obviously a lot more durable, but also more expensive and more permanent.

We've had good success laying down unused wooden pallets and filling the gaps in with gravel and dirt in our goat shed that had bad drainage. This elevated the floor 4 inches off the bare dirt and made sure that water wouldn't sit and collect. With plenty of fresh straw bedding, our goats have a dry, comfy place to sleep that will stay out of the rain.

Of course, you don't have to use anything if you don't want to. A concrete slab is preferred by some, especially for smaller buildings that you'd be able to lay yourself, and is not only easier to clean but will prevent animals from digging in under the walls. Gravel, dirt, or straw are other common floorings for sheds, greenhouses, chicken coops, and other outbuildings.

Building Material

Most outbuildings on the homestead are made of wood, and with good reason. The price is reasonable, materials can sometimes be sourced free of charge if you have a chainsaw and splitter, and it's easy to work with. Wood has some insulating properties as well, which is nice for things like chicken coops, shelters, and greenhouses where you would want to keep winter cold out.

Sheets of metal are common for barns and larger buildings because after the wooden frame is built the metal sheets are a quick weatherproof siding to be applied. Metal sheets are often made of tin—we call it "barn tin" around here. The barn tin will last a long time, although rust will eventually weather it. The thin metal sides are less insulating from the cold and we found them unsuitable for use as a chicken coop so we converted the large tin-sided building to a garden shed.

Some homesteaders get very creative with small outbuildings. I've seen sheds out of homemade adobe bricks. I've seen rubber tires stacked into walls, packed with mud, and made into a goat shelter. Sand bags were stacked into the dug-out side of a hill for a root cellar. Natural rocks that were moved out of the garden area were used to create the foundation for a three-sided shelter for sheep and goats. One trait of the backyard farmer is the ability to frugally use whatever is on hand to get the job done.

Size and Situation

Yes, it's a bit of a hassle to build an outbuilding. And it is more of a hassle to build a larger 8×10 building compared to a smaller 4×4. However, it is an even bigger hassle to build a small building, use it for a month, get frustrated, tear it down, and build a second, larger building because the first wouldn't do the job. As I've mentioned, you'll want to make sure the size of your outbuilding is appropriate to the job for which you intend to use the building. It is also a huge hassle to try to move a building after you've built it, especially if you are digging the corner posts into the ground for stability. So it is important to make sure that wherever you build your shed, coop, or green-house is where you really want it.

Make sure that there is plenty of clearance around the door area to get in and out easily. Make sure the building won't block the sun for your garden vegetables that are trying to soak up the rays. You also want to consider the weather for buildings that are used to house animals. If your area has cold winter winds from the north, don't situate the door on the north side of the building where drafts will blow in around the door.

Leave space between buildings and fences so you can walk all the way around the building if needed. Keeping a clear path around the building, especially chicken coops, will help minimize (though not eliminate) predator attempts to devour your flock.

Which brings me to my final point: make sure the building you end up with will do the job you need it to do. Chicken coops, for example, will probably need extra reinforcement around the walls and floors to keep predators from digging into the coop. Ventilation screens are needed to allow good airflow, especially in the summer. A garden shed should have enough space to hold all the tools you want to store.

Fences can be built out of a wide range of materials as well. On our homestead alone we have hog panel, chain link fence, privacy vinyl fencing, and chicken wire fences. Each has a different purpose and use, although some could be interchangeable.

We fenced in our garden area to protect the plants from the ravages of our crazy dogs, kids, and goats. For this fence, we only need something tall enough to deter the goats, and squares small enough to contain the chickens when we let them roam through the garden space. So we use 1×2-inch welded wire square fence but purchase the 5-foot-tall wire instead of the more common 4-foot-tall wire.

If deer are rampant in your area, even a 6-foot fence may not be enough. You can double-fence your garden to help deter deer by placing a second 6-foot fence that is 4 feet from the perimeter of the first fence. This "moat" area in between the two fences is too wide to be jumped in a single bound and will deter deer from attempting to enter your garden. I know one gardener who uses that 4-foot-wide area around the outside of the garden as a secondary chicken run.

For the goat pen and side yard run (used as extra grazing area or in case we need to shut the dogs up in a smaller space), we use cattle panels or hog panels. These large welded wire panels are much more substantial than what's used around the garden, and they come in 15-foot lengths. They are easy to tie onto metal stakes driven into the ground at regular intervals, and they stand 5 feet tall. The stiffer wire holds up against abuse from the goats standing against them, trying to reach their heads through, etc.

In a pinch, we can lash a panel into a circle for a makeshift hay rack, or move a panel and tie it across a corner to separate a goat that is in quarantine or if we think she might be sick. This year we're even using the fence panels as tomato trellises in the garden, so we've found these to be extremely flexible.

We call these panels "cattle panels." Here you can see they are made of sturdy wire that holds up to extra weight. When my in-laws switched to electric fence we used these panels for our goat pens.

The back fence between our backyard and the neighbor's yard is chain link because that's what we had on hand at the time. Nothing goes to waste, and it was easy to finish the yard with the chain link that we had available. Chain link is also very flexible and often used in suburban homes, but the typical 4-foot height would not be sufficient to hold a goat. Plus, chain link fencing can be more expensive than other livestock fencing options mentioned. Price what is available in your area before spending a lot of money on fencing, and make sure the fence you buy is suitable for the job.

When you are adding fences to your property, you want to allow grazing and foraging animals as much access to grass and brush as possible. Sometimes, this means it is better to fence in the garden instead of fencing in your animals. If you have room, consider having more than one grazing area for your animals so you can let one section grow grass while they eat down another area.

Never build a fence without considering your gates. I find misplaced gates to be highly annoying. And we have one in our backyard that drives me crazy! It is awkwardly placed so you have to go out of your way to check the goats or get into the pen for milking. You can be sure I'm plotting a better design for later. Learn from my mistake and plan the gates so they line up with your primary paths, walkways, and logical flow of traffic!

Also make sure your gates are large enough for whatever needs to go through them. It's great that your cute little garden gate is big enough for you to get through. But can you get your wheelbarrow through? Likewise, your goat pen—if your goats eat round bales you will need to be able to get that round bale of hay into your goat pen because, trust me, those things are heavy if you are trying to roll them across your backyard.

It's not hard to adapt your current situation into a fruitful, bountiful backyard farm. Think creatively about using the space available to you. Make every plant count. Give each bit of land a purpose. Be intentional and purposeful about how you move forward in working with your home. I can't wait to see what you come up with!

Gardening on a Backyard Farm 2

This part is all about the earth and growing things—everything from creating a good foundation for a healthy garden, to saving seeds and other advanced gardening techniques. See how to start out a garden the right way, with respect for the land and an eye toward longevity and sustainability.

You'll also be introduced to some of the most popular and useful plants in the backyard farm. Fruits, vegetables, herbs, and edible perennials are discussed, with tips for growing each and a list of some common varieties you might want to try. I also show you how to maximize your garden production so you can grow more food than you thought possible in whatever space you have available.

Learn the benefits of heirloom and open-pollinated plants for economic savings, increased independence, and a healthier genetic diversity. When you're ready to try your hand at seed-saving, you'll learn the basic methods for success.

What and How Much Do You Want to Grow?  4

Getting your garden set in place is an amazing feeling. It is literally putting roots into your community and your home. Transforming part of your backyard into a productive place that will actually feed your family is brilliant—if it works. And for that to happen you want to make sure you keep a few things in mind when you plan and build your garden, including where your garden should be and what plants to grow.

You'll not only need to plan your garden, but also determine the best style of garden for your situation. Whether you use traditional gardening methods, raised beds, or containers, you'll want to gather the needed materials ahead of time.

Planning Your Garden Space

First you need to think about what your family eats. This is an area where I think a lot of hobby gardeners fail. They focus solely on a limited number of vegetables, which generally make up a small portion of their typical diet. When trying to support a family with your food, you want to think about nutrient-rich, high-protein, and high-caloric plants such as fruits, potatoes, beans, and grains.

My husband loves tomatoes, so we devote a significant portion of our garden to tomatoes. We eat a lot of fresh tomatoes while they are growing and I preserve them by making sun-dried tomatoes and salsa, which are things we use a lot in our family. I can get organic pasta sauce for a reasonable price, but salsa and sun-dried tomatoes are more costly to purchase in our grocery store, so that's what I focus on. Watermelon, raspberries, onions, asparagus, green beans, and squash are other items we eat a lot, so we find lots of space for these plants in our garden.

I also look at the cost of getting organic or locally grown produce in my area when working with a limited space. I've found that herbs in particular are often overlooked by the average home gardener, and fresh herbs are usually very pricy to purchase. For example, my local grocery store charges more than $3 for a small package of fresh basil, but a packet of seeds for basil is only $1. I can grow $50 worth of basil (or more!) in just one summer season.

You can grow a variety of fresh herbs in a simple and inexpensive container garden.

It's really hard to give specific numbers for people who are beginning gardening. I wish I could make your planning easier by telling you that if you use 1,000 square feet for your garden area you'll be able to grow 50 pounds of vegetables, but that's not possible. I would be wary of holding too fast to any list that tries to give you exact numbers when it comes to food production. What I've found is that not only does that number change year to year, but even which crops do well changes from year to year.

One year our tomatoes did not do very well, but we had a lot of peppers, squash, melons, and cucumbers. This past summer we didn't get peas, but our tomatoes and green beans did particularly well. There will always seem to be some ebbs and cycles in the garden. Weather, seed fertility, disease, and a variety of other factors can all contribute to which crops will do well in any given year.

OVER THE GARDEN FENCE

Keep a record of which varieties of plants do well in your area, and save seeds from the plants that perform the best in your garden—you'll see your productivity level increase over the years.

Square-Foot Style Gardening

One of my favorite methods of gardening, especially for beginners, is the Square Foot Gardening method. While I don't follow this method completely, I do use a lot of the principles recommended as a starting place for my garden each year. Mel Bartholomew is the creator of the Square Foot Gardening method and author of *All New Square Foot Gardening*. I'm going to briefly discuss the principles of Square Foot Gardening and what aspects can be adapted to various kinds of gardens.

The Square Foot Gardening method breaks up the garden into 4×4-foot sections with paths set in between each of the large squares. Each square is then broken up into individual square-foot sections and planted according to the number of plants allowed to occupy each space. Seeds are sown singly to avoid overcrowding and soil is amended as the crops are planted and harvested. With this method gardeners can produce a large amount of produce in a more condensed space than industrial farms. They also cut back on weeding and watering because of how the plants grow close together and help shade out weeds.

A square-foot garden is the first garden I planted when I moved to Oklahoma with an 8-week-old baby in August 2006. We didn't have a lot of time left in the gardening season but I wanted to do something to make it feel like home so I dug up a 4-foot square and put in some overpriced tomato starts and some fall-season seeds like lettuce and spinach.

This square-foot garden has been divided into individual squares. Within each square a single plant or type of plant is added to the bed. The squares on the left are planted with tomatoes, while the two squares on the right side are planted with Verbena (bottom) and strawberries (second from bottom). When Serene adds other plants such as carrots, radishes, or bush beans, she will plant more than one in each of the single squares because those plants can be spaced closer together.
(Photo courtesy of Serene Vannoy)

Even after I outgrew this gardening method in our own backyard garden, I still use the close spacing techniques. I also adopted the trellising of my vegetables after finding it so successful and now grow as much as possible vertically. This lets me pack as many plants as possible into our garden spot—we have to fence it to keep the critters out, which means I really am working with a finite space.

Raised-Bed Gardening

Raised beds can be used to create garden spaces without having to dig into the ground. Essentially they are huge containers, boxes without the bottoms. I've seen raised beds as short as 2 inches and as tall as 3 feet. They are really useful when building gardening spaces in less-than-ideal conditions. If you have a lot of concrete, hard clay, or infertile soil to deal with, or just want to work your garden without stooping and bending over, a raised bed might be a great solution.

Raised beds are built with wood, brick, or stone walls in most of the versions I've seen. Railroad ties are popular in my area and provide a wide, sturdy side wall. Regular lumber like 2×4 boards or even repurposed boards from around the house can also be used. Natural stone is available inexpensively for those who live in rocky areas but will require manpower to dig up and place.

If you are building your raised beds on concrete, you'll want to build them up high enough so that the soil inside the raised beds is able to support the entire root system of your plants. This means at least 18 to 24 inches for most plants and more for perennials such as berries and asparagus. If your raised beds will be sitting over soil, you can probably make them shorter because ultimately the roots will reach into the soil below. The poorer your soil quality is, the deeper you'll want to make your raised bed; however, if you amend the soil you can improve the quality of the soil below while still enjoying an immediate harvest. You'd also need to make the beds deep enough to choke out any weeds from the ground or lawn below.

One way to inexpensively fill the bottom of your raised bed while improving the soil below is to use the lasagna gardening technique. Lasagna gardening uses layers to improve the soil and smother weeds and traditionally involves very low-key ingredients. Start with a thick layer of newspaper (the black-and-white pages, not the glossy pages) that is at least several sheets thick so it will smother out all the grass and weeds. Then layer in soil builders like compost, straw, manure, sawdust (from nontreated wood), leaves, grass clippings, and other biodegradable plant materials. You can make an entire raised bed of this mixture and let it age a season before planting, or bring in topsoil for the uppermost levels and get started planting right away.

The biggest benefit of raised beds is that you're able to control the quality of the soil for your plants. Your soil can be filled with organic material and won't be compacted or hard for roots to stretch. Raised beds also have excellent drainage, which is handy in places with heavy rainfalls or if your land is high in clay and has poor drainage.

The downside of using raised beds is that they are more costly at the start. That initial outlay of expenses is more than just your seeds and plants, but also includes lumber to create the bed, and soil to fill the bed. In some areas the cost of purchasing topsoil can be expensive, so take that into account when considering using a raised bed for your garden. On the other hand, if you can collect leaves, pine needles, and grass clippings, you can get a good start on filling the raised bed with compost.

OVER THE GARDEN FENCE

Some composting materials can change the pH level of the soil. Pine needles, for example, can create a more acid level in the soil, especially if used in large amounts. Using a variety of materials to build your soil can help keep things balanced. (See Chapter 6 for more about soil.)

Traditional Garden Plots with an Intensive Twist

A traditional ground-level garden plot can grow a tremendous amount of produce for your family, if you ignore the typical growing techniques that have been shrunk down from industrial farming. The typical farm with hundreds of acres of land in production uses wide rows designed for mechanized planting, watering, and harvesting. Instead of trying to translate the monoculture and mechanical methods of large-scale operations into the backyard farm, grow more with less using intensive and Earth-friendly methods.

Our vegetable garden is naturally raised a little with all the layers of compost and mulch we've added, but we don't have it built up with timber like a raised bed, or limited to 4-foot by 4-foot sections like the Square Foot Gardening method. We do use a lot of the techniques found in both these methods to grow our garden with an intensive twist you just can't duplicate without the hands-on attention and care of a small backyard farm.

For example, we plant our seeds close together not just down the row, but across multiple rows. So if bush beans are to be planted 4 inches apart, I will plant down a single row and space the seeds 4 inches apart. Then I step over 4 more inches and plant a second row with bush beans. We're able to reach through the plants for harvesting without any trouble, and get a lot more plants into our garden that way. I'll usually plant about 3 feet before leaving enough space to walk through the garden.

We also seed the garden very carefully. I don't scatter seed and then thin the plants that sprout up. I plant each seed individually so plants are already spaced nicely. This means I'm not only controlling how close together my plants grow, but I'm also saving a lot of money because I'm using all the seeds I plant.

Planting our potatoes closely allowed the plants to act as a living mulch, shading out weeds. We planted these rows 3 feet across.

Intensive gardening methods have several benefits for the backyard farmer:

- ⚘ Planting in large swaths of land means you can concentrate all your soil-building efforts and know it will be used by the plants. Instead of laying down compost over the entire field, and then leaving dirt unused between every row, you are making the most of the fertile soil by planting the entire area.

- ⚘ You can water the garden without wasting your water on areas that aren't producing anything. The water conservation efforts of an intensive garden will be greatly improved with plants shading the dirt beneath them and every drop of water falling where it will be used.

- ⚘ Weeding is easier with intensive gardening because the plants help shade out and block out weed seeds. Weeds that do begin to grow are easily plucked out.

- ⚘ Growing plants vertically, up the trellis or fence, can make a huge difference by giving you space in an extra dimension. I discuss vertical gardening in detail later in this chapter.

Keep a Garden Journal

Garden planning is something that never ends. Every year we try something new or stop doing something that didn't work so well the year before. Now we are growing our plants from seeds and saving them year by year.

Your garden journal, or garden log, should be a useful tool you add to each year and refer to on a regular basis. Some of the things you'll want to make note of in your journal include these:

✢ A planting diagram of your garden layout and what went where each year. You think you won't forget but the truth is, you probably will. Especially after more than a single year.

✢ A planting log of what you planted, when you planted it, when it sprouted, when you transplanted it, and how it performed. This can include the germination rates of each variety and how it did in the garden through the year.

✢ A list of where you purchased your seeds in case the germination rates varied from one company to another.

✢ Notes about pests or diseases you encountered through the year. Especially important if you are saving seed!

✢ How much you harvested from your vegetable garden of each type of crop.

✢ Do's and don'ts for the following year. You always think you'll remember but it's better to write things down.

✢ A list of monthly chores performed or to perform. If you plan to sell your produce or livestock, you will also want to track your financial expenses.

✢ Note water and fertilizer usage in your garden to see where improvements can be made the following year.

See Appendix B for some garden journal sheets that you can print out and use. These will help you both in planning your first year's garden and in improving your garden each year thereafter. You'll be able to track which crops and varieties do best in your region. You can print out additional copies at BackyardFarmingGuide.com.

Maximizing Your Garden Space

The other way to maximize your garden's produce is to use your space wisely. Small-space gardening techniques can be used in the backyard farm to make sure that you are getting the most from your land—however large or small that may be.

Whether you have a quarter-acre like my husband and me, or a small suburban backyard, you can still grow a great deal of food for your family by implementing the three space-saving techniques discussed in this section: vertical gardening, container gardening, and edible landscaping.

Vertical Gardening

One of my favorite ways to save space and grow more in the backyard garden is to grow vertically. The traditional method of growing many plants is a space-consuming technique and can limit the amount, or even types, of crops grown.

By using the vertical gardening space, you'll be able to grow more produce. And you'll be able to include some of the larger vegetables and melons that you might not think to include in a home garden. The following is a list of produce that are well suited to grow vertically:

Blackberries	Peas
Cantaloupes	Pole beans
Cucumbers	Pumpkins
Gourds	Raspberries
Grapes	Watermelons
Indeterminate tomatoes	Zucchini

There are so many ways to take advantage of the vertical space in your backyard homestead: trellises, arbors, fences, hanging plants, cages, teepees, and more! The options are hugely varied, but we'll touch on a few of the most common and easiest to implement.

Trellis. One of my favorite options for growing vegetables vertically is using a trellis. In our current garden, we have a permanent trellis structure that we use for a variety of vegetables depending on the current crop rotation. One year we might plant it with melons and cucumbers, and the next year we'll train our tomatoes up the trellis.

In a previous garden space, I used a rope trellis support for our tomatoes and tied it at an angle from the backyard fence. The ropes were a one-season-only solution, of course, and would have needed retying for the next growing season, but they worked beautifully.

When I'm trellising melons (and squash) in my garden I like to stick with melons that mature at around 10 pounds or lighter so I don't have to tie the individual fruits with netting of any kind. Also keep in mind that because of the more perennial nature of fruits like blackberries, raspberries, and especially grapes, you may need to use a more permanent trellis material such as wire rather than rope.

We secured the rope at the top of our solid privacy fence and staked it into the ground. These indeterminate tomatoes grew well up the slanted rope trellis.

Arbor. A permanent structure like an arbor is a classic solution for growing grapes. I've also seen adorable vegetable garden entrances with gourds and cucumbers growing over the top of an arbor.

Fence. Don't ignore the already-existing structures in your backyard. Fences can be a great asset for backyard farming. You can create a trellis on a solid fence, as I did with the tomatoes in the previous photo. You can also use the fence itself as a trellis when you have something like hog panel, chain link, or other open fencing types. Do keep in mind, when you're using fencing as a trellis, that you'll need to have a fencing material sturdy enough to support the plants.

Cage. These are a staple of the classic vegetable garden, most often seen with tomatoes. Traditional two-tier cages tend to work best with *determinate* tomatoes versus *indeterminate* tomato varieties. See Chapter 8 for details about growing tomatoes in the backyard farm.

DEFINITION

Determinate tomatoes are typically bush form plants that reach a specific height (say 4 to 5 feet tall), then flower and produce fruit all at once. Determinates should not be pruned by removing the suckers at the stems or fruit will be diminished. These are best suited for cages and fruit is often used for canning, paste, or sauce making in large batches. **Indeterminate** tomatoes are typically vining plants that grow continuously throughout the season, flower sporadically, and produce fruit steadily until first frost. They can grow as tall as 12 feet or more and will dwarf (or topple) a traditional tomato cage. Indeterminates are usually considered more tasty and flavorful.

Teepee. Like trellises and arbors, a teepee provides support for climbing vines. A cane pole teepee planted with several bean or pea vines is a classic garden icon for good reason. The space saved by growing these climbers upright is space you can use to tuck in more plants.

Garden teepees made of bamboo can be a great way to grow beans, peas, or cucumbers vertically in a small-space garden.
(Illustration by Becky Bayne)

Hanging plant. Vegetables that can be grown *up* can also be grown *down*. Create more space for your vegetables by planting them where they will hang down. A great example of this is the bags that grow tomatoes upside down. Just make sure they get plenty of sun and water.

Container Gardening

Container gardening is a popular way for small-space gardeners to enjoy the fun of gardening when they don't have a yard. But there's no reason that container gardening has to be limited to ornamental plants! Use containers whenever you need to eke a little extra growing space out of your backyard farm.

Vegetables grown in containers do need some extra attention in order to produce well. Keep these basics in mind:

- ⚘ **Watering:** You'll need to water a container more often. Unless you use a self-watering container, plan to check your container crops on a daily basis.

✦ **Soil quality:** Containers are notorious for losing the micronutrients out of the soil after repeated watering. Be prepared to add compost, fertilizer, and mulch on a regular basis—especially for heavy-feeding crops like tomatoes.

✦ **Room to grow:** Generally speaking, the bigger the container, the better. A larger container will allow crops a chance to mature at a normal pace without being stunted. If you must use a smaller container, choose dwarf varieties.

I love using containers for my herb garden so I can keep them close to the kitchen. There's nothing as wonderful as working on a dish, stepping out the door, and harvesting fresh basil or thyme to add to that night's dinner. And for me, the easier things are to manage, the better.

I see people stressing out about what makes a good container or not. The truth is, anything can become a container. I've seen strawberries growing in an old bath tub that someone repurposed from a landfill. The main requirements for a container when growing edible plants are the ability to hold enough soil and good drainage. That's it! Beyond that, the sky is the limit.

We've used wooden barrels, EarthBox self-watering containers, old horse watering troughs with rusted-out bottoms, children's wading pools that cracked and would no longer hold water, repurposed nursery pots that landscaping plants came in, and more. Don't feel like you have to spend money on "real" containers from the garden center. Backyard farmers are notorious for not spending money unless they absolutely have to.

Trailing rosemary in a glazed, terra-cotta container.

THORNY MATTERS

Most of the crops in a backyard farm can be adapted to a container. However, there are some that don't do well at all. Asparagus, for example, is a perennial vegetable with a huge root system that wouldn't be sustainable in a container.

Check the mature size of the plant for an idea of whether you can expect it to do well in a container. Don't forget that you can add a trellis or cage to a container and even grow climbing plants such as cucumbers or squash. The following plants are examples of specific varieties that are ideally suited for container life.

Tomatoes:

Balcony Tomato. As the name implies, this red cherry tomato plant grows to a dwarf 2 feet in size that won't overrun a medium container.

Gold Nugget. One-inch, round tomatoes are a cheerful golden color on bushy plants that grow about 3 feet tall and wide.

Orange Blossom. A true orange hybrid tomato, this variety produces 6- to 7-ounce tomatoes.

Valley Girl. Eight-ounce, red tomatoes that are tasty, and the plants tolerate temperature variations, making it hardy in a container.

Eggplant:

Fairy Tale. Only growing to about 18 inches, the purple fruit comes in clusters that reach about 5 inches in length.

Dancer. Medium fruits that reach 8 inches long, this eggplant is a compact but high-yielding variety.

Squash:

Bush Pink Banana. Only 3-foot-long vines produce several large winter squash fruits. You'll need to trellis them but not as far as most squash vines.

Carrots:

Atlas. Only 2 inches long, this is one carrot you could grow in a container without impeding the root size.

Lettuce:

Bibb. An heirloom lettuce with delicate flavor that grows only 6 to 12 inches.

Buttercrunch. An All-America winner that forms tight heads of 6 to 10 inches.

Herbs:

Greek Dwarf Basil. This is a great compact herb plant that only grows 6 inches tall.

English Munstead Lavender. A favorite of mine for containers, lavender can tolerate the drier conditions of life in a container.

Fruit:

Sunshine Blueberry. Only 3 feet tall and wide, this is a self-pollinating blueberry, making it ideal for a container garden.

When trying to determine if a particular variety will do well in a container, look for hints in the product description. Words like "compact," "bush form," or "nonvining" are good clues.

PLANT STANDS

Elevating your containers can allow you to add even more room for growing produce. Don't be afraid to lift a container out from behind a shady wall so it will get the six to eight hours of sun it needs. Or use a porch rail or window sill for an herb garden, placing plants at eye (and nose) level where they can best be appreciated.

Edible Landscaping

Another fun way to add more produce to your overall production each year is to consider the ways to substitute edible plants into your ornamental landscaping. Like a harvest-producing ninja, you'll sneak a row of rainbow chard in a sidewalk border and use thyme as a ground cover along your walkway. Suddenly, in all the places you used to grow plants that only looked good, you'll have plants that can ultimately feed your family (or your livestock) instead!

The main thing you need to pay attention to when making this switch is to consider the gardening basics I cover in Chapter 6, such as how much sunlight the area receives. Here are some suggestions for how to make easy switches for edible landscapes:

- ✥ **Ground cover:** Instead of sedum, creeping phlox, or vinca ground covers in your ornamental beds, plant thyme or lettuce.
- ✥ **Flowering perennials:** Instead of daisy and *Coreopsis,* plant chamomile and echinacea.
- ✥ **Ornamental annuals:** Instead of impatiens, pelargoniums, and petunias, try Swiss chard, basil, and cilantro.
- ✥ **Climbing vines:** Instead of morning glories, honeysuckle, or wisteria, try planting peas, pole beans, small squash, and grape vines.
- ✥ **Landscaping shrubs:** Instead of one of the many large flowering shrubs, try elderberry, shrub roses, and blueberry plants.

- ❧ **Ornamental trees:** Small trees like crab apples and ornamental pears can be replaced with true fruiting counterparts such as peaches, pears, apples, and citrus. Dwarf forms can even be grown in large containers!

- ❧ **Shade trees:** Traditional shade trees like catalpa, magnolia, and oak can be switched for nut-bearing alternatives such as pecan and walnut, or full-sized fruit trees.

When you are looking for ways to make the switch, think about what aspect of the ornamental plant you appreciate the most. Is it fragrance you enjoy? Try rosemary, old-fashioned roses, or elderberry. Is it colorful fruit that you appreciate? Consider persimmon, blackberries, or other edible fruits. Perhaps the foliage has a unique color like gray or purple? Plant lavender or tri-color sage.

Note the use of beautiful pepper plants in this urban raised bed. The foliage is as pleasing as any other bedding plant, and the fruits are highly ornamental as well.
(Photo courtesy of Baker Creek Seeds)

Whatever your gardening situation, chances are there will be ways to include more edible plants into your traditional landscape design.

By using ninja-planting tactics, placing containers in otherwise unplantable situations, and managing your crops like a professional, you'll find that even the smallest backyard farm can produce an abundance of food.

Tools and Skills for the Backyard Farmer  5

Maintaining a backyard farm can be an excellent way to not only increase the health of your family, but also save money, as long as you already have some tools and you are prepared to do some or all of the work yourself. If you purchase everything brand new, or have contractors come out to do all the work for you, you will not save money. Not even close.

Just as there are some tools that should be part of every backyard farm, there are some skills that every homesteader will want to know (or develop). The savings can be astronomical.

For example, the price of a small, well-built chicken coop at our local market was $450. The cost of my husband repurposing some old lumber and creating a chicken coop? Only $50 worth of chicken wire to cover the outer yard area. That's $400 in savings right off the top. (See Chapter 1 for a photo of the finished coop.) If someone else builds your fence for you, it can cost anywhere from $10 to $15 per hour for a full day's work, in addition to the building supplies and cost of materials.

If you are paying for everything out of pocket, chances are you're losing money. Whether it's worth it to you in terms of the cost versus time outlay to have the increased health benefits is something only you can decide for yourself. However, it is still necessary to have a basic understanding of repair skills because you have to expect the worst to happen, at the worst times. If your goat breaks through the fence in the middle of the night, guess when you have to fix the fence? And if you think you can call the local handyman for help at 10 P.M. you are sadly mistaken.

This chapter covers the basic skills and tools you should have or try to pick up before tackling your backyard farm.

Hand Tools

There are certain hand tools that are absolutely necessary. I love a good spade, for example, and unlike my husband I do almost all my gardening using simple hand tools. He is the king of the shovel and triangle cultivator but I like to pick a cozy spot and hunker down and feel the dirt when I work.

OVER THE GARDEN FENCE

When it comes to garden tools, size definitely matters. Pay attention to the handle size and test different lengths to see which is more comfortable for you, because if the handle is too short your hand tools will cost you a lot of energy to use. I like to find the handles that are in the 18-inch range for my hand tools, while my long-handled shovels and hoes are better in the 36- to 40-inch range given my shorter stature.

Can you have too many hand tools? Not as far as I'm concerned!
(Photo courtesy of Brannan Sirratt)

Garden Spade or Trowel

A garden spade is a must. I actually have two different styles that I like to use for different purposes. Plus my kids can confiscate one and I still have another to work with. Really, though, while they can be used interchangeably, there are times when one shines over the other.

A narrow spade is easier for creating small holes for seeds, or digging in dry, compact soil. The narrow blade on this spade means I won't disturb the roots of other plants in an established garden bed, mixed border, or container planting.

I also have a deeper garden trowel that will move a larger amount of dirt. This is perfect when I'm transplanting plants and want to preserve a larger rootball. It's also good for digging a slightly larger hole, a longer row, or building a new container planting, as it will move dirt faster.

Pay close attention to the make of the tool before you purchase it. Cheap wooden handles will dry up and become loose very quickly, so look for high-quality ash handles. Or consider the soft-grip, comfort handles with strong resin or vinyl handles. Weak metal blades will bend with the kind of use these tools will get at the backyard farm. High-carbon steel blades are more durable than flimsy metals and will hold up to lots of use.

Secateurs or Pruners

These handy tools are a must for the backyard farmer. You'll definitely want a handheld version of pruners because it will be a lot easier to do controlled cuts on a fruit tree, berry bramble, or rose bush. There's a level of control you can get with handheld tools that is really nice, and if you work through your garden on a routine basis you'll be able to snip off branches while they are still young and easy to cut.

Bypass secateurs give a cleaner cut because instead of crushing the branch, the sharp blades are brought past each other. Be sure to buy a pair of pruners with blades that are easy to clean and sharpen, because with repeated cuts the blades will build up resin and sap residue and become dull. Some high-end pairs of pruners have gear-type mechanisms that help close the pruners, making them a better choice for gardeners with a weak grip.

Look for quality metal, not pressed steel, or blades that can be replaced when old ones are no longer serviceable. Handles should be comfortable to grip, sized appropriately to your hand, and made of a quality hardwood, like ash, or durable material.

OVER THE GARDEN FENCE

Just as it's worth paying more for a good-quality metal, it's also worth paying more for strong handles on your tools. You want your garden tools to last, so durable handles are a must. Quality hardwoods are one of the most popular materials for their longevity and durability.

Digging Fork

Digging forks are a must. They are extremely useful for breaking up the hardpan layer on new garden beds. After double digging a garden bed, before I add the bottommost layer of dirt back into the trench, I like to use a digging fork to break up the dirt as deep as possible. The strong, long tines of a digging fork allow you to punch deeper into the soil, loosening compacted areas so roots can delve deep for nutrients and water without as much effort.

Long-handled digging forks are also useful for turning over a compost bin. A digging fork's open design allows less resistance than a shovel when trying to dig into a matted or compacted area. As with all garden tools, quality hardwood handles or durable nonslip grip handles will last much longer. Tines of the fork should be forged metal, not easily bent or broken.

I prefer short-handled digging forks for lifting established plants, as they will help preserve more of the root structure. I prefer long-handled digging forks that can be used when standing, for breaking up the bottom of digging holes, and for turning over compost bins. They can also take the place of a rake and shovel combination when cleaning out the muck from the goat pen or chicken house.

OVER THE GARDEN FENCE

Digging forks can be used to harvest root crops like potatoes because the tines are less sharp than a spade and are less likely to damage the vegetables during the harvest.

The shovel has a full blade on it and is primarily used for digging or moving dirt. The digging fork has tines that are strong and sturdy and can be used to break up hard soil, aerate compost heaps, and many other garden chores.
(Photo courtesy of Christiane Marshall)

Hand Weeder and Hoe

A handheld weeder is a must in an established garden area. A full-size hoe could easily chop an established desired plant in a crowded vegetable bed or front border. A handheld weeder lets you get close and control your deadly blows to only the undesirables.

A new favorite type of hand weeder is the cobrahead weeder and cultivator. The curved head makes it comfortable to use and the longer digging portion of the tool means you can dig out long-rooted weeds like thistle or dandelion. The fact that the head is very thin allows you to get into small spaces between established plants.

The cobrahead cultivator makes weeding easy and lets you get in next to other plants already growing.
(Photo courtesy of Baker Creek Seeds)

One of my husband's favorite weeders is the push-style, draw, or scuffle hoe. These have heads that are parallel to the ground and can be pushed along the ground just under the surface of the soil, severing the top of the weed from the root. We have a triangle-head weeding hoe that fits easily between growing plants. It's used a lot at the beginning of the season when grass weedlings (what we call weed seedlings) are just starting to sprout and the vegetable transplants aren't yet large enough to shade them out.

A stirrup hoe is another style that is usually used in a push-pull motion just under the surface of the soil. Again, it works to separate the leafy parts of the weeds from the roots, and is most effective before the weeds are well established. By chopping off their heads when they are still baby weedlings, you break up their growing cycle and starve the weeds out until your healthy veggies grow big enough to block the sunlight. Just picture yourself as the Queen of Hearts and run through your garden shouting, "Off with their heads!" and you'll be alright.

Traditional hoes are great for digging furrows if you are planting crops in rows. They can be used to move mulch out of the way or hack a stubborn weed root. We have a wider-head hoe that we use both early in the season, and later in the season to cut up spent vegetable plants for adding to the compost bin. We also have a thin, angled hoe that my husband uses for miscellaneous garden chores.

Sometimes we select a particular tool for a specific area or job and sometimes we just grab whatever is close at hand. Regardless, you want your garden tools to have quality metal heads that won't rust, bend, or resist sharpening when needed. And you want your handles to be the right size (shorter if you're shorter and longer if you're tall) and a high-quality material that is shaped comfortably for your hand.

Cultivation

While most people think of cultivating the land with a giant tractor, a smaller garden space means less expensive equipment! Whether you choose to use a mechanical device like a tiller or do everything by hand with a good shovel, a good-quality tool is a must.

Tiller

Tillers are used to dig into the ground and break it up into loose soil suitable for planting. In new garden areas, a tiller will cut through small roots, grass mats, and turf, as well as turn over the soil so you can get rid of rocks. A tiller comes with either rear tines or front tines. Rear-tine tillers are easier to maneuver and turn than front-tine tillers. Many tillers are self-propelling, which means that the tines help move the tiller across the ground.

We have a medium tiller with deeper tines that we use in tandem with double digging. We also use the tiller to turn under green mulch (I'll discuss growing cover crops in Chapter 6) and increase the fertility of the soil. Depending on the size of your garden areas a tiller may not be necessary, but we get a good amount of use of our tiller in our backyard. You won't need a large tiller regardless.

THORNY MATTERS

Before you invest in an expensive tiller, find out if there is a small machine shop in your area that can work on them. Inquire at the shop which brands they are able to get replacement parts for and which brands they can't work on. If you can't have a simple item fixed on the tiller you buy, you're stuck with a useless machine that would otherwise need only a minor repair. They might also be able to tell you which brands they see a lot of—a hint to avoid them!

Shovels and Spades

These are the primary must-have gardening tools, and for most small-space gardens a shovel is the only tiller you'll ever need. In many situations you can avoid the expense of a tiller completely (ours was a hand-me-down) and do all the garden preparation with a good, well-built shovel. Of all the tools you'll invest in, the shovel should be the very last place you try to cut corners or save a buck.

Round-point shovels are the most common and most versatile shovels available. Every gardener needs a round-point shovel with a long handle to give them some leverage. Remember to sharpen your shovel as soon as you get it home because shovels and other gardening tools are always sold dull. Choose a blade with enough angle that when the blade is flat on the ground, the handle should be at a proper angle. This angle between the head and the handle of the shovel is called the lift.

Test the lift you need by setting the shovel at a straight angle so the head is straight, not angled as most people dig. You want your arms to be fully extended in front of you so when you pull your arms to your body, you will get the maximum lift, and use the least amount of work. If you have to push the handle past your body toward the ground you'll strain your back and tire yourself out faster. Check out the video at BackyardFarmingGuide.com for a demonstration of good digging technique.

As a woman, I should mention that it is important to find a shovel that is also the right height for your body. I like the D handle shovels both to get the proper lift and to be able to grip the end of the handle without overbalancing myself. I have to be picky about the shovel I choose! I also use my lower body to push the shovel in, so it's important to me that the step plate, or top edge of the shovel that you can mash on with your foot for extra oomph, is wide and well formed, instead of skinny and painful to step on. The bottom line is that you want your shovel to be comfortable because you'll be using it. A lot.

Transplanting shovels, like transplanting spades, are long and thin shovels designed to get into more crowded growing areas. This is a tool that when you need it, proves its value on the first use by saving your other plants growing near the plant you're trying to dig up. If you don't have a transplanting shovel, a garden fork might work well instead.

A square shovel, or cutting spade, is rectangle shaped. It's often used in landscaping to create the clean-cut edges between planter beds and lawns. We tend to use it as a muck shovel because the flat blade surface will scrape up the manure off the floor of the goat pen or chicken house more easily. For that alone it will save you enough time and trouble to make it worth the purchase. Again, a comfortable grip is an absolute must because soggy hay and manure can be heavy by the shovelful.

Garden Rake

This isn't your springy, wimpy, autumn leaf rake. This is a true garden rake. Forged metal with tines that are stiff instead of yielding, the garden rake is heavy but moves easily over fresh-tilled soil. Nothing is better for creating a nice seedbed for planting small seeds than a good garden rake. You can push it deeper into the soil to loosen dirt clods and fish out rocks, or you can float it over the surface to create a fine seedbed for outdoor sowing.

The handle should be long enough to extend your reach across the full width of your garden bed (we have our garden sections 4 feet wide) without bending over or overreaching. Ours is actually completely metal with a metal handle and rake both making it balanced to maneuver, if heavy to actually lift and carry around. The heaviness of the rake means it will also move wet or soiled hay from the goat pen and rake it into a pile more easily. A leaf rake would just skip over the top of the muck and not actually move it into a pile.

Strong metal tines and a heavy head mean less force is needed to dig out dirt clods and rocks.
(Photo courtesy of Brannan Sirratt)

THORNY MATTERS

All tools with angles like a rake, spade, and shovel should be placed carefully when not in use. You know the cartoons where the gardener steps on the rake and it smacks him in the face? Don't be that guy.

Fencing and Miscellaneous Homestead Tools

There are a lot of other tasks around the backyard farm other than just gardening, and basic building and fence building seem to be never-ending. These tools are items you'll use again and again to build or repair your fences and outbuildings.

Posthole Digger

This awkward-looking contraption is used to dig a deep hole straight into the ground at any kind of depth. A posthole digger has two handles and a pincher mouth at the end with two moveable pieces. You press the handles together, which forces the mouth pieces apart, and drive them into the ground with force, then pull the handles apart, pinching the jaws shut, which will close up and hold a chunk of sod and dirt. Lift the jaws (don't let them come apart and drop your dirt!) and move the dirt to the side, closing the handles back together as you do to open the jaws and release the dirt. This process repeats until the hole is as deep as needed for your fence post. Most fence posts are set at a depth of 2 feet. That's really deep to try to dig with a shovel, so a posthole digger is a must.

Don't try to dig a posthole when the ground is too wet, or your hole will fill with water.

We use a posthole digger to set clothesline posts, fence posts, and even corner posts for sheds and stationary outbuildings. I've also used the posthole digger for deep-rooted asparagus crowns because some of the roots are 2 feet or longer. It is handy for planting deeply planted bulbs in a pinch, too, and when working as a team, my husband Sidney and I planted 100 bulbs in an hour. He'd grab a pinch of soil, lift it up, I'd place in the bulb, and he'd replace the plug of sod right back where he grabbed it from.

It would be a lot harder to set a post without a posthole digger.

Make sure the moveable pieces of the jaws are well made. Cheap metal pins will break from the strain and leave you with two useless halves. The handles must be comfortable to use or you'll soon tire out. Be sure that your blade edges are sharp! I cannot stress this enough. File your posthole digger edges after every use. It takes just a minute, and you'll always have a fresh, sharp edge.

THORNY MATTERS

Always check for the placement of underground wires, gas lines, or water lines before digging around in your backyard. A quick call to your utility companies can save a great deal of hassle—or even your life.

Fence Pliers

These are especially handy when working with wire-and-post fencing as the pliers are specifically designed to hold wire (barbed or unbarbed), pull staples, cut wire, and assist in twisting wire so it stays in place. Fencing pliers are crazy looking—my kids call them dragon pliers—but they can sure get the job done. Good fence pliers also have a hammer piece and enough heft to act as a hammer when needed. It's a great all-in-one tool for fence work, maneuvering chicken wire cages, or repairing holes in a panel or welded wire fence.

Wire Cutters

These won't need to be a separate purchase if you've picked up a true pair of fencing pliers, which will include wire-cutting capabilities. To make your own anything on the homestead you'll need a good pair of wire cutters. You can snip pieces of baling wire to use as ties on your trellis in the garden. Wire a gate closed. Make a compost bin out of scrap lumber and chicken wire scraps. Put together a brood box. Cut a length of welded wire that's too long for the area you're trying to section off. Wire cutters can cut any metal pieces of wire and fencing.

A good grip is a must because these tools rely so much on hand strength. Be sure the handles aren't so far apart you lose your leverage, and look for high-quality metal blades that won't dull too quickly. It's a pain to have to stop halfway through a project to sharpen your blades because they won't hold an edge.

Wheelbarrow

You must have a good wheelbarrow for the garden and backyard farm. Most of us won't need a mechanical garden cart or tractor in our small, intensive spaces. But a wheelbarrow will get a lot of use, and a high-quality model will last for years if cared for properly.

Look for a sturdy wheel instead of a cheap tire with tubing that will go flat within a couple months. When it comes to durability and ease of use, a wide tire is so much better, as a thin, hard tire can sink into soft soil or soggy areas, making it ten times more difficult to move your load. Check the wheel bearings carefully and look for a high-quality construction that won't break down.

On the plus side, a two-wheeled wheelbarrow is more stable and less likely to tip over while moving. It can also be easier to push over shifting ground like loose gravel or sandy paths. On the downside, a two-wheeled wheelbarrow makes wider turns and needs more room to move through a garden area. When maneuverability is more important than stability, a one-wheeled wheelbarrow might be best.

A two-wheeled wheelbarrow will be more stable than its one-wheeled counterpart, but harder to turn and maneuver.
(Photo courtesy of Brannan Sirratt)

If you get a metal wheelbarrow bucket, or tray, be prepared to repaint it on a yearly basis to avoid rusting out the bottom and to cover the scratches that will inevitably develop during use. If you choose plastic, be sure that you select a UV-resistant plastic that will withstand rotting in the sun so you'll get more than a couple years of use out of your wheelbarrow investment. Either material is fine but whichever you choose, select a model with a deeper tray so your load won't bounce or slide out. A deeper tray will also allow you to put the heaviest portion of your load over the wheel, making it easier to move.

Test the handles before you purchase a wheelbarrow and make sure the grip is easy to hold. Our first wheelbarrow had sturdy handles but they were so thick they hurt my hands to grasp for any length of time. You also want the handles to rest close enough to the ground when the barrow is still that you can easily lift the wheelbarrow up for maneuvering around the yard.

Check the struts and braces to make sure they are made of strong metal that won't bend under heavy loads. Any metal pieces should be made of rust-resistant metals or painted to avoid rusting. Clean any manure or debris out of the wheelbarrow tray, and off the metal struts, to avoid encouraging rust, and if you store the wheelbarrow outside don't let rainwater collect inside.

STORING YOUR WHEELBARROW

Store your wheelbarrow upright so the metal stands, or braces, aren't in contact with the wet ground to build up moisture and corrode. Be sure that water isn't left to stand in the tray of the wheelbarrow, as this can rot the bolts in the bottom of the tray, even with plastic wheelbarrows. Plastic wheelbarrows should be stored out of direct sunlight to avoid warping and weathering. Space-saving hooks and braces are available so wheelbarrows can be hung on a shed or garage wall up and out of the way.

Your wheelbarrow is the single best way to move heavy or wet loads such as chopped wood, gravel, rocks, paving stones, compost, manure, bedding, or topsoil any distance in your backyard, so be sure you choose one that will hold up. It's also good for giving the kids a fun ride on a lazy Saturday afternoon.

Other Must-Have Tools

Other basic tools that you'll want to have handy are items you probably already have around the house. These are all tools we use on a weekly basis, for some project or another:

- Hammer
- Screwdrivers (Phillips head and flathead)
- Pliers
- Safety goggles and face mask
- Tape measures in different lengths (a 50-foot tape measure is handy for outdoors)
- Sturdy stepladder
- File (for sharpening all those tools we talked about earlier)

- Cordless drill
- Circular saw
- Sawhorses
- Crowbar
- Square
- Level
- C-clamps

We've found that picking up quality tools, and taking care of those tools, has saved us a great deal of money in the long run. The cheapest tools are usually the hardest to work with and the quickest to break. The frustration factor alone makes it worth spending a little more money up front.

A variety of tools are needed for regular building jobs and tasks around the backyard farm.

Caring for Your Tools

Care for your tools by cleaning the blades after every use. Once a year give them a thorough treatment with a stiff wire brush, and sharpen the edges at this time. Check any bolts, pins, braces, or joints to make sure they are tight, not rusting, and clean from debris. When they are clean, spray with WD40 to prevent rust during the winter.

Treat hardwood handles once a year as part of your regular winterizing checklist by sanding them lightly, and then oiling the handles with linseed oil. Store your hand tools blade-down in a bucket of sand with enough motor oil added in to make the sand slightly wet. The linseed oil keeps the handles water resistant, and the sand/motor oil treatment keeps the blades rust resistant. Now your entire set of tools will last much longer and work better whenever you're ready to head to the garden.

> **OVER THE GARDEN FENCE**
>
> The motor oil in the sand bucket doesn't need to be new oil. Most people I know who use this trick will save their used motor oil from an oil change on their cars. Country repurposing at its finest!

Basic Building Skills

I've tried to cover some of the how-to-use information when I was discussing the various tools for the backyard farm. And of course, more information about specific uses will be found in the chapter discussions where that information is most needed. But this section briefly touches on some of the basic building skills that are needed for keeping a backyard farm profitable. You can't afford to hire out simple things like making a compost bin or brood box (see Appendix A for a simple plan to make your own brood box).

The best way to learn basic building skills is to work alongside someone who is more experienced than you are. I learned to use power tools in my theatre classes at the university where all students were expected to help build set pieces. I made it a point to learn each tool and find out the principles of building the various types of set pieces, and I'm grateful for that knowledge now. There's not much difference between cutting a 2×4 to build a square set wall piece and building a compost bin, except where you place the lumber.

Cutting Lumber

A basic circular saw, mentioned earlier in the chapter as a must-have tool, can be used for a lot of the lumber-cutting needs on a backyard farm. Combined with a pair of sawhorses to elevate the lumber, you can cut lumber, posts, and plywood as needed to assemble any number of sheds, outbuildings, cold frames, or other structures.

These homemade sawhorses, made from scrap lumber, make it easier to cut your own lumber at home.

Sidney used to have an old hand-me-down circular saw that had been at his family's place forever, I think. Once we had kids I insisted that we invest in a newer saw with all the modern safety features and guards in place. Look for a quality saw that has a finger guard to protect yourself from accidental cuts, and a safety lock feature that will prevent the saw from accidentally turning on. These functions provide greater safety when using your saw.

THORNY MATTERS

Always wear safety goggles to protect your eyes from flying sawdust and chips of wood. If you're cutting sheetrock or painted lumber, wear a face mask as well to protect your lungs from inhaling particulates from the air. Watch out that you don't accidentally cut the power cord and be sure the area where the cut lumber piece will fall is free from children, pets, or feet.

Mark your lumber for cuts carefully! Nothing is worse than the feeling you get when you place that board up and it's the wrong size. It's such a waste that could be avoided by carefully double-checking your measurements prior to making the actual cuts.

When you're marking a cutline on a short piece of wood like a 2×4 or fence post, you can use a square or level and simply draw a straight line end-to-end in the appropriate place. For a long board like a sheet of plywood, the marks can be a little more complicated. Your square won't be long enough to allow you to draw straight down the edge with a pencil.

To mark a board like this, measure one side and make an arrow or notch drawing with the point at the exact spot to cut. Then measure the other side of the board and make a similar mark. Use a chalk line stretched taut between the two marks, lift it up slightly, and let it go to "pop" it so a straight chalk line is created. This will be your straight cutline to follow when you are cutting your board. Sidney says if you don't have a chalk line you can use a 2×4 but the line might not be as straight.

To actually make your cut, place the board securely on the sawhorses with the cutline in the empty space between the horses (you don't want to accidentally cut your sawhorses!). Lift the guard out of the way, and place the exposed blade near the board but not touching. Then hold your saw firmly with two hands and turn the saw on.

Press it forward slowly as the blade cuts the board allowing you to advance the saw forward. Watch the line! Keep your mark in line with the saw blade so your cut will be accurate and straight.

When you reach the end of the board, the feel of resistance in the board and the sound will change— slow down and be prepared for the board to fall when the cut is completed. If you have a partner helping you, he or she can hold the piece of the board you're cutting (if there's room to safely hold the board without getting too close to the saw) and keep it from falling. Release the trigger as soon as the board is cut to stop the saw blade from spinning and prevent any accidental injuries.

Putting the Wood Together

We often use screws to join our wood pieces as it's easier to assemble, and in the case of mistakes, disassemble, as compared to nails. In windy weather, or extreme heat that dries out the wood, nails can be prone to loosening and coming out of the wood. Screws are less likely to do so. Backyard farmers with an eye on the budget will find that nails are less expensive than screws, however, so it's up to you which method you want to use and when.

If you are screwing two pieces of lumber together, you will find that a high-quality (a.k.a. not the cheapest) cordless drill makes your job a lot easier. Invest in a drill that allows you to reverse directions easily and has interchangeable heads for a variety of tasks.

OVER THE GARDEN FENCE

You should have more than one battery for your cordless drill, and always have a charged battery on hand. Nothing is more frustrating than getting halfway through a project and having to stop to charge your drill.

When fixing lumber together at a right angle to form the corner of a box or wall frame, use more than one screw or nail. Multiple screws will stabilize the boards. You can usually screw from the outside piece so that the points of the screws are not exposed where you can get cut. If in doubt, screw (or nail) from the thinnest piece of wood into the thickest piece of wood as when attaching a plywood siding piece to a 2×4 frame.

Line up the end of one board with the side of the other board to get a good idea of where you'll be attaching the boards. Don't finish tightening your screws, though, until you've used your level to make sure everything is as square and level as possible.

If you are screwing the boards together near the end of the board, or attaching thinner pieces as with framing or trim, you will probably want to create a hole in the wood with your drill first. These are called pilot holes or clearance holes, and they allow the screw to easily penetrate the wood without splitting, going off course, or leaving a gap between the two pieces of wood. Use a bit that's slightly smaller than your screw when creating these pilot holes so your screw still has something to grab on to.

Always use the right bit for the job. Never use a bit that is worn or has nicks in it, because it won't turn the screws properly. Sidney prefers star head screws as these won't slip off the screw as easily. Phillips head screws tend to strip off more easily, and are more difficult to back out if needed. He also uses a magnetic bit holder, usually an extra purchase, but worth every cent because it helps prevent losing your bits.

Hold your drill at a strong right angle and press firmly. If the screw doesn't go in easily, you may need to drill a pilot hole first.

Hold your drill at a right angle, use firm pressure, and advance the screw evenly. If the screw goes all the way in, but the two pieces of wood haven't tightened together, back the screw out of the second piece, press the two boards firmly together, and then screw it in again. When using small screws, especially brass or aluminum screws, change the clutch on your drill to the lowest setting to avoid snapping off the screw head.

Most building projects are an adaptation of basic building skills. It's not any different to build a greenhouse than it is to build a shed or a chicken coop except that the materials used may be slightly different, and the measurements will change. You will find that investing in even the most basic building skills will pay huge dividends in what you save on the backyard farm.

Gardening Smart from the Start   6

When you grow fruits and vegetables, you are growing plants that are usually greatly changed from their wild ancestors through selective breeding by humans who wanted more and tastier fruits. They are less hardy and less tenacious but bear larger and more delicious fruit. As a result, they need the gardener to tend them as they've lost much of their wild-survival adaptability. You give them what they need to thrive, ensure their future generation, and in return get delicious, nutritious fruits and vegetables.

This means you have to understand the basic needs of your garden produce. There are basic requirements for the plants you're trying to grow that have to be provided for to ensure the maximum harvest.

Soil: A Dirty Word

The literal foundation of your garden is the soil. You've surely begun to grasp by now how I obsess about never taking more from your soil than you put into it, and the reason for that is simple. In a backyard farm, the health of your soil quite directly affects the health of your family. If your soil is lacking in nutrients, your diet could be negatively affected. Plus, rich soil will grow healthier, more productive plants.

Healthy soil is alive with millions of microorganisms in a single square foot of garden space. It's loose in texture, easy for plants to send their roots through. The organic matter that's contained within ensures that the soil doesn't get compacted so hard that no air can reach the roots of your plants. Healthy soil is like a well-tuned piano—it doesn't become tuned by accident and it doesn't hold a perfect pitch forever.

Types of Soil

It may seem strange, but all soils except peat-based soils started out as rocks. The differences in particle size are partially what are responsible for the different types of soils and the way they behave. *Clay soils* have tiny rock particles, hold water for a long time, and don't allow a lot of air into the soil. *Sandy soils* have large rock particles, with lots of air able to get into the ground, but the water drains away very quickly and so do the nutrients in the soil.

There are three main types of soils that are found in the garden: the two extremes of clay and sand, and the ideal soil, what many gardeners call *loam*. Loamy soil is neither sand nor clay, but rather a mix of each with plenty of organic matter mixed in. Unfortunately, most backyards don't magically have this type of soil already present.

There are other types of soil you might have in your area, although they are less common.

Peat soils form in wet, acidic areas and the soil is usually dark, almost black. The soil doesn't hold its shape and retains moisture for a long time, but, like clay, when it dries it can become very dry.

Saline soils tend to have a high *pH,* can be infertile, and are usually found in arid climates.

DEFINITION

The **pH** of the soil is the acidity or alkalinity of the soil. The pH of your soil affects how available the nutrients are to be used by your plants. The ideal range of pH for most vegetables is between 6 and 7.

You can tell a lot about your soil composition by giving it the squeeze test. When your soil is slightly damp, like a wrung-out sponge, pick up a handful and give it a firm squeeze in your hand. Heavy clay soil will retain a lot of water and will stick together like a sausage in a solid, slimy roll. Highly sandy soil tends to dry out quickly and the clump will break together easily into very small pieces when you open your hand. The more you can avoid either extreme, the better.

The tiny mineral rock particles aren't the only ingredients in the soil. It's also organic matter, microorganisms, air, and water. The exact makeup of your soil determines the pH and the fertility of your soil. In every case, the best way to improve the quality of the soil and make it better for your home garden is to add soil amendments such as compost, peat moss, leaf mold, and other organic matter.

Feeding the Soil

Bottom line: the better your soil, the better your garden. I used to think that this meant mixing up some of that pricy, chemical fertilizer and pouring it over my plants on a regular basis when watering. I've since learned that overuse of synthetic fertilizers can actually damage my plants, wash into the water supply, and cause deficiencies in micronutrients that aren't supplied. The answer is to build up your soil with organic methods like composting and nonharmful fertilizer.

Composting to Build Healthy Soil

Adding compost is one of most efficient ways to feed your soil and improve the soil quality. And it's not something you do once; feeding your soil is an ongoing process that is part of your regular maintenance in the backyard garden.

I prefer compost over chemical fertilizers because of the huge number of benefits from composting:

- ♣ Economically, the cheapest soil-builders are the ones from your own garden, livestock, and kitchen. Why throw it out if you can compost it and add it to your garden for no money at all?

- ♣ From an ecological perspective, avoiding the petroleum-based fertilizers is an obvious benefit.

- ♣ Building your soil with compost gives longer-lasting benefits because chemical fertilizers, while quick and easy to apply, are usually water-soluble and wash away very quickly.

- ♣ Soil that is rich in organic material will hold heat better, helping gardeners extend the growing season.

Making compost is not rocket science. It's earth science. In its simplest form you throw a bunch of biodegradable stuff together in a pile, and a few months later you have fabulous dirt to add to your garden. Remember when I said it's okay to be a get-it-doneist instead of feeling like you have to be a purist? This is one of those areas, because if trying to do compost "perfectly" means you don't do anything, then stick with a heap-it-in-a-pile method. Having said that, there are a few things you can do to increase the effectiveness of your compost pile.

The Right Ratio

The key is to make a compost pile that cooks or matures—that is, breaks down from your raw materials into *humus* more quickly. Finished compost is almost always a 30:1 ratio of carbon to nitrogen. The microbes that work to break down the materials in your compost pile live and work best in a ratio of 25:1 to 35:1, so providing a mix of high-nitrogen (called *green*) materials and high-carbon (called *brown*) materials is the best way to build up your compost bin quickly.

DEFINITION

Humus is the final product of the breakdown of organic matter. It is dark, woodsy smelling, and full of nutrients that benefit both soil and plants.

Now don't get overwhelmed by all that. The key to remember is that when you put a high-nitrogen item like, say, chicken manure (10:1) into your compost bin, you would want to also mix in a low-nitrogen ingredient such as chopped leaves (50 to 80:1). As I said, this isn't an exact science but rather a principle to keep in mind when working in the backyard. If you have mucked the goat pen and the chicken pen into the compost bin that week, you might decide to throw in some dry straw, shredded paper, sawdust, or even cardboard.

The following table lists some common green and brown matter you can find readily in most households and backyard farms.

Green Matter (High Nitrogen)	Brown Matter (High Carbon)
Fresh vegetable scraps (15:1)	Straw or aged hay (80:1)
Grass clippings (20:1)	Sawdust (500:1)
Tea bags/coffee grounds (20:1)	Wood chips (400:1)
Animal manure (10 to 20:1)	Leaves (chopped or shredded) (60:1)
Seaweed and algae (19:1)	Cardboard (egg cartons/toilet paper tubes) (350:1)
Weeds or plant prunings (20:1)	Shredded paper (170:1)
Human and animal hair (10:1)	Corn stalks (60:1)
Wood ashes (25:1)	Wheat straw/oat straw (60:1)

We like to keep two main compost heaps going—one that is just finishing and one that is in the process of being used. We pile everything we can into one, as quickly as possible usually. I empty my paper shredder and we give the animal houses a thorough cleaning to get the new pile started. From there it's just a matter of adding to the pile during our regular household chores. When you start thinking about the compost bin in the backyard, you find quite a lot of what usually ends up in the landfill can be repurposed to your backyard farm!

Keep a bucket dedicated to composting around the house and add your scraps to it throughout the day. In the evening when you close up the chicken coop, empty the bucket into the compost bin.
(Photo courtesy of Tim Sackton)

Mowing the lawn? Toss the clippings into the compost pile and mix them in. Weeding? Toss it in! Our coffee grounds and kitchen scraps and shredded cardboard paper tubes all go in a bucket on the counter and then out to the compost bin each day. Whenever I add anything that might mat up, like a bunch of grass clippings or leaves, I am careful to turn the pile over and stir it up a little bit. You want air to be able to get inside the compost bin so the whole thing heats up and breaks down. By the time the bin is full, you've got a good mix of raw materials added in.

Now let that compost bin sit and age and let the microscopic workers do their thing and start filling the second compost bin. By the time that bin is full, your first one will have cooked down into dark, rich, fertile compost ready for you to add to your garden. If you find that you are filling your bins faster than the compost is becoming ready (this might happen if you have lots of manure or household waste you're adding, as our large family does) then simply add another bin.

ON A DIFFERENT SCALE

Have you heard about how the Native Americans would bury a fish where they planted their corn, peas, and squash? That was something we've now fancied up with the term "trench-composting." You can do the same thing in your own garden by digging a deep hole, filling it up with vegetable scraps from the kitchen, and then covering it up with dirt. Plant heavy feeders like squash or melon vines on top, and voilà! You've composted with no hassle or overthinking.

A couple of quick words about some special composting ingredients:

> ⚜ Wood chips and other hard materials like corn cobs or avocado pits won't break down in the compost bin well unless you shred them up in a wood chipper, so most gardeners avoid these altogether.

> ⚜ Coffee grounds will raise the acidity of the soil in large quantities, so some people prefer to age their coffee grounds in a separate compost pile reserved for acidic-soil plants like blueberries, camellias, and hydrangeas.

> ⚜ Wood ashes are taboo with some gardeners because they raise the pH levels in high amounts, but I like to add them for the micronutrients (namely potash) and because we have a wood-burning stove so it seems wasteful not to compost something that could be composted. Wood ash shouldn't be a problem unless your soil already has a high pH level or if you add a lot of them into the mix.

If your compost bin begins to stink, you've probably added too many large leaves or a big mound of grass clippings that have matted together into a soggy, slimy mess. To solve this problem, add in some bulky material like straw and stir up the entire pile. Your compost bin should be kept moist, but not soggy, and you want to make sure that it gets plenty of air into the middle of the pile.

This homemade compost bin is constructed from leftover lumber and old chicken wire. It allows plenty of air flow in the compost pile.

Moisture, air, and a proper mix of carbon and nitrogen will give the compost microbes a perfect environment to work in and your compost bin will heat up, killing any weed seeds and plant diseases that may have been lurking on the ingredients you added.

Consider the Source

It is important to consider where your compost materials come from. I know gardeners whose vegetable plots have been ruined because of the residual herbicides in the manure of cows that were not fed organically and sustainably. If you are sourcing your manure from animals that are being wormed, given growth hormones on a regular basis, or are eating hay that has been sprayed with chemical herbicides, all of that can find its way into your family's vegetables.

Additionally, grass clippings from lawns that have been treated with herbicides and pesticides, or sawdust from treated lumber, can act in the same way—transmitting unwanted chemicals to your vegetable garden. It's another example of how the agricultural industrial complex has taken something that was designed to work in perfect balance (matter feeds soil feeds plants becomes matter feeds soil) and so disrupted the original cycle that it's now toxic in places. That's why I tend to err on the side of caution and only use compost materials from our own backyard farm, or my in-laws' ranch.

The more organic matter you can prevent leaving your backyard farm to the landfill, and remaining to feed your land, the better off your soil—and your produce—will be.

Organic Fertilizers and Soil Supplements

It seems that all farmers have a fertilizer blend or soil amendment they just swear by, and I think that's because soil varies so much in composition and makeup. Our area in Oklahoma tends to benefit from small applications of lime, but other areas have different deficiencies to contend with.

OVER THE GARDEN FENCE

I highly recommend taking soil samples to your local county extension office for testing. You'll learn if there are any specific deficiencies that you need to address. Deficiencies in your soil become deficiencies in your crop, which in turn become deficiencies in your family's diet!

Even with healthy soil as a base, I like to add some plant food to our heavy feeders like asparagus, tomatoes, and potatoes. I found a great recipe for organic fertilizer in the book *Gardening When It Counts: Growing Food in Hard Times* by Steve Solomon. The mix is inexpensive when made in bulk and can be applied right onto the soil around the plants.

Commercially purchased soil amendments often come in large plastic sacks.
(Photo courtesy of Brannan Sirratt)

Of course, there are lots of things that can be added to your garden as soil amendments. These organic fertilizers are from mineral, plant, or animal origins and usually contain many micronutrients and trace elements in addition to whatever main nutrients are in them.

Bonemeal. This supplement is made from finely or coarsely ground bones, usually slaughterhouse remnants, and is an excellent source of phosphorus and calcium. It usually encourages strong root growth, which is why you'll hear gardeners say to mix a spoonful into the bottom of a planting hole for bulbs. It also helps prevent blossom end rot in tomatoes.

Bloodmeal. Like bonemeal, this fertilizer has an animal origin and is powdered blood. Over 10 percent nitrogen, it is a strong fertilizer that is useful to feed nitrogen-loving plants and leafy greens. Be aware that spread bare in the garden, bloodmeal might attract dogs or cats to the garden.

Gypsum. A common ingredient in plaster and construction drywalls, this fertilizer contains calcium sulfate and helps supply calcium without changing the pH of the soil. You can use it to lighten clay soils, often in tandem with dolomitic limestone, but it isn't one of the most effective or popular fertilizers as not every area needs gypsum.

Feed meal. Usually soybean meal or cottonseed meal, these fertilizers are high in nitrogen (6 to 7 percent) and potassium or phosphorus. They can be used as a side dressing on vegetable beds to feed fast-growing annuals. Of interest to those looking for organic backyard farms, cotton crops are often heavily sprayed with pesticides and most soybeans are genetically modified.

Alfalfa meal. Alfalfa is a great source of not only nitrogen, but many other nutrients as well, such as potassium. It's often used as a general fertilizer for feeding both annuals and perennials, but many self-sufficient farmers are beginning to grow alfalfa as a cover crop to improve soil fertility without having to apply it in concentrated form.

Seaweed meal or kelp. Valued for the diversity of minerals and nutrients it provides, gardeners can usually find it in liquid, pellet, or powdered form. Applied to soil in the spring and summer, seaweed applications seem to increase a plant's stress tolerance. It is considered by some to be the most complete source of micronutrients available.

Rock phosphate. This is a nice source of phosphate for vegans who want to avoid bonemeal. However, it is stronger—especially hard-rock phosphate—and it shouldn't be used on alkaline soil because it will raise soil pH. Soft-rock phosphate is not as strong and breaks down more slowly so that it lasts for a long time in the garden.

Limestone. Both ground limestone and dolomitic limestone raise a soil's pH level and supply calcium. Dolomitic limestone also supplies magnesium and is easy to spread directly onto the soil.

Zones and Plant Hardiness

Knowing your hardiness zone (the number assigned to a region according to the coldest expected temperatures) will help you determine which plants will grow best in your area. These numbers are especially important for growing fruit trees and perennial herbs that will need to survive a winter in your area. The U.S. Department of Agriculture (USDA) hardiness zone map is a general guideline of the United States that helps gardeners get a basic idea of the weather conditions in their area. The smaller the number on the hardiness chart, the colder the winter temperatures.

For example, my growing zone in southeast Oklahoma is 7b, which means we can expect temperatures with a low of 5°F to 10°F over the winter. This means that an apple tree that is hardy to zone 5 would probably survive a winter in my area, while an orange that is hardy to zone 9 would get too cold and die. But this is only half the picture.

The other zone map that is handy for gardeners to consider is the heat zone map by the American Horticultural Society. This map tracks the average number of days that are above 85°F. So my rural Oklahoma corner of the world is a heat zone 8, with between 90 and 120 days in the heat zone (above 85°F) each year.

These two numbers taken together start to give you an idea of the types of plants that are ideally suited to your general area. There are several ways to manipulate your specific backyard, however, to try to include plants that might be outside your zone. The best way to do this is to pay attention to the *microclimates* in your backyard garden and learn how the layout of your land will affect the temperature of a small, specific area of your yard. Have you ever noticed that after a snowstorm, some parts of your yard are completely free from snow very quickly, while other areas still have unmelted snow a week later? Those are microclimates at work, and you can use them to your advantage as a gardener to grow as wide a variety of plants as possible.

Are you trying to grow an apple variety that needs more chilly hours than you are zoned for? You could choose a lower-lying spot to plant your apple tree, as the bottom of a slope can be a frost pocket where the cool air gathers. Does your area have higher temperatures than your lettuce and broccoli prefer? Try planting your lettuce in an area west of the cucumber trellis so it is provided diffuse shade for part of the day, creating your own microclimate.

In general, southern exposures get more sun and tend to be warmer, while northern exposures get less sun and tend to be cooler. Areas that are elevated tend to be warmer, while the low depressions can allow cold air to collect. The exception would be at the top of a hill where cold wind exposure makes it easier for plants to freeze. Providing a windbreak can raise the temperature for an area by creating an insulating buffer. I'll cover ways to get more out of the garden, regardless of your gardening zones, in Chapter 7.

Sun and Water

Another of the vital components to a healthy garden is sunlight. Each vegetable and fruit plant has a natural cycle based on the number of hours of sunlight it receives. Most vegetables need a full six to eight hours of sunlight to produce well. However, the following vegetable plants and herbs will tolerate fewer hours of direct sunlight, thriving even if they only receive four to six hours:

Broccoli	Dill
Cabbage	Greens
Calendula	Lettuce
Chard	Mint
Cilantro	

Of course, the more sun a plant receives, the more water it is likely to need. And water is another area where conservation on the backyard farm can really make a big impact. Everything that I've talked about so far can impact how you water your garden. Of course, my focus is on making it easy, while not wasting water unnecessarily.

It is best to water deeply and thoroughly, so the water soaks completely into the soil and encourages the plant's roots to grow deep into the ground. Loose, friable soil makes it easy for plants to grow strong, deep root systems like this. Creating healthy soil with plenty of compost and organic material worked into the garden soil will also help conserve water because those organic particles act as little sponges, holding the water in the soil where the plants can access it when it's needed.

THE SCRATCH TEST

I stressed out about watering the garden when I first started until my father-in-law, with all his country wisdom, set me straight. "It's easy," he said. "Just water the garden when it needs to be watered." He taught me how to tell when it was time to water by scratching into the dirt in the garden a little bit and if the soil was not moist just below the surface, it was time to water again.

Now most vegetables can tolerate a little bit of watering irregularity but they all taste better and grow better without that added stress. Keeping plants evenly moist will help them grow more successfully and can, as in the case of tomatoes, make the produce healthier. I like to achieve this the lazy way by using a soaker hose and a thick layer of mulch. In the spring after I've planted the bulk of my long-term seedlings I lay a soaker hose through the garden bed. When the soil has warmed up, I add a thick layer of mulch as well (adding mulch too soon can insulate cold soil and keep the ground cooler, longer) to prevent evaporation from wind and sun exposure.

By using these two techniques in tandem I can use the smallest amount of water possible, applied directly to the plants that need it, and make that water last as long as possible. Granted, last summer with triple-digit heat for three months in a row, we were still watering every day, but we used thick layers of mulch to help prevent drought stress and keep many of our vegetables going throughout the entire summer. Don't forget that watering in the hottest part of the day will waste more water than watering in the evening or early morning. We often set our timer to water between 3 A.M. and 4 A.M. so the plants will be fully watered before the sun rises.

Organic and Land-Friendly Principles

You've noticed that so much of my focus so far has been about the land. The soil is the foundation of your entire farm. It is important to create a system of balance where we provide for the land and the land provides for us. There are two more ways of keeping the land healthy I want to touch on that will help improve the long-term viability of your small-scale farm.

Mulching your garden is almost like composting right in the garden bed—if you use plant-based mulches, of course, which I usually do. There are so many benefits to mulching your garden that it's almost insane to not place a good mulch cover on at the end of spring. Here are some reasons to mulch a garden:

- ❦ Mulch helps prevent water loss. Much of the water lost in the garden is lost through evaporation from wind and sun. A thick layer of mulch helps prevent that by providing an insulating layer.

- ❦ A generous layer of mulch is excellent weed prevention because it limits the amount of sunlight that reaches the weed seeds. Weeds that do sprout are easier to pull out.

- ❧ Plant-based mulches can feed the soil over time as they break down. While the mulch doesn't get mixed into the soil like compost does, there is a small benefit for your plants.

- ❧ Mulch helps moderate soil temperature so that your plants are cooler in the heat of summer, and given some protection from frost in the late fall.

- ❧ A mulched garden usually looks more attractive than an unmulched garden, especially with plant-based mulches.

Now, before we discuss a few of the many items you can use as a mulch, there are a few things that you should *not* do with mulch. Laying a thick layer of mulch over a cold ground in the early spring can actually prevent the soil from warming up, so wait until the ground warms up a little bit before laying down your mulch. The exception to that would be black plastic mulch, which can help heat up the ground.

Applying mulch directly onto the stem of the plant, or trunk of the tree, can encourage the buildup of moisture and invite diseases. To avoid this, just pile up the mulch under the leaves of the plant so it isn't touching the stem of the plant.

There are so many materials, even if you only limit yourself to plant-based and biodegradable materials, which can be used as mulches. And each has pros and cons.

Straw or hay. This is what I like to use because I have access to a free source. Hay can bring in weed seeds, but I've found that laying down at least 4 inches of mulch (and planting our vegetables closer together) can help prevent most of the seeds from germinating. Hay allows air and water to easily reach the soil and is easy to rake out of the way when needed. We use hay that isn't suitable for feeding and is either peeled off the outside of the round bale, or has been discarded at the bottom of the feeder after being picked over by the goats.

Straw makes an excellent organic mulch option. Make sure the source of your hay or straw hasn't been treated with pesticides.
(Photo courtesy of Tim Sackton)

Wood chips or bark chips. Wood chips, or bark, is a very attractive mulch and lasts a long time in the garden. If you like to till your garden each year, you might not want bark chips because they will often last longer than a single year. Some types of bark, like pine, can slightly change the pH of the soil. I've also read warnings that a deep layer of bark mulch can encourage rodents, but with our barn cat and our dogs, that hasn't been a problem I've ever experienced.

Grass cuttings. Grass cuttings are readily available to most households (just put a bag on your lawn mower and collect them!). However, grass clippings can't be applied thick enough to suppress weeds without forming a wet, slimy mass, and they can also introduce weed seeds into the garden. On the plus side, grass cuttings biodegrade quickly and provide nitrogen to the garden when they do.

Chopped-up leaves. Another easy-to-find mulch, leaves work well to suppress weeds. Again, some types of leaves, like oak, can lower the pH in the garden soil. If you don't chop the leaves up they can form a mat, similar to grass clippings, so run them over with the lawn mower before applying to the garden.

Leaves from the fall can become an excellent garden mulch. It is recommended to chop them up before laying down in the garden.
(Photo courtesy of Brannan Sirratt)

Pine needles. For some people, pine needles are free for the taking, and look attractive in an informal way. Pine needles, like oak, are a little acidic. I like to add them around my camellia bush and use them as mulch around other acidic-loving plants, like blueberries and hydrangeas. They do a good job of preventing weeds while allowing air and water to reach the soil.

Cocoa shell hulls or buckwheat hulls. I have never used this mulch, or seen it used in this area, because it is costly to purchase. It seems wasteful to buy what is otherwise available so inexpensively from other sources. Cocoa shell hulls are very attractive, though, and long-lasting because they don't break down as fast as mulches like hay or chopped leaves. Buckwheat hulls are considered very attractive and preserve moisture well, but if allowed to dry completely, the mulch can blow away in high-wind areas.

Black plastic. This material is often used by commercial growers because it heats the soil quickly, blocks weeds, and lasts a long time. But for organic farmers, black plastic can cause a dilemma. If you are trying to minimize your impact on the environment, do you really want to use a mulch that will end up in a landfill instead of feeding the soil underneath it? Besides that, the one time we tried plastic we had to stake it down when the strong Oklahoma wind blew it around, and then weeds grew through the holes, defeating the purpose.

Using cover crops, sometimes called green mulches, is another way to keep your land in good health. This concept of letting the land lay fallow, or not producing a crop that is harvested but rather planting it with a cover crop, is something that is mentioned as far back as the early books of the Bible! In between planting vegetables, you can rotate in a leguminous cover crop like clover, cowpea, lentils, or hairy vetch.

There are a few things to keep in mind when planting cover crops:

- ⚜ Time the planting of your cover crops with the season. Winter crops like oats and winter peas (often planted in a pair) or hairy vetch can be started in the late fall and allowed to work through the winter.

- ⚜ When it's time to replant with something else, you'll need to mow or cut down the cover crop. Sometimes that's just pulling out the cover crop and throwing it in the compost bin, or running a cutting hoe underneath the ground an inch to separate the plant from the root. (For us that means turning the chickens loose to work through that specific bed, where they do a perfect job ripping up the ground, tilling things over, and preparing the bed for replanting.)

- ⚜ Many cover crops will become pests in the garden if you let them set and scatter seed before you remove them. This tendency is lessened when you use chickens to till them up because chickens will eat many of the seeds.

The main thing about cover crops is to choose something well-suited to your specific area (check with your county extension office for personalized tips) to keep your soil from being bare in between plantings. Bare soil becomes hard-packed, suffers nutrient loss, loses top soil, and doesn't provide anything of value to the backyard farm. Using a cover crop in between plantings helps keep your garden areas producing at the highest levels.

Get the Most from Your Garden  7

One of the first things I thought when we moved to a large city lot was that I wouldn't be able to have a large, idyllic garden. I had a picture in my head of a big field planted with rows and rows of crops. What I got was about a quarter-acre total, so our personal home garden is only 20×60. What can you grow in a small space? As it turns out, you can grow a lot!

My husband Sidney's parents live on 40 acres, and their garden last year was more than 2 acres. But our small plot out-produced theirs by the end of the 2011 growing season.

In this chapter, we look at all the tools backyard farmers have in their arsenal for maximizing their garden's potential. Whether you have a few containers on a backyard patio, or a larger garden plot to work with, you can grow more food than you would think at first glance.

Some of these techniques will be the basic steps that everyone can take right from the start. For example, vertical gardening is as simple as creating a trellis and tying up your vegetable plants as they grow. Others might seem more advanced at first, but trust me, if I can do them, so can you!

Increasing Crops Through Intensive Gardening

Planning my garden is the best part of winter for me. I love to doodle in my garden journal, paw through pretty garden catalogs, and dream of warmer weather. What I've learned is that by planning your garden wisely you can double the amount of produce you grow each year.

Using crop rotation and planting in succession with multiple plantings in a single year, you'll be able to grow more produce in less space. Yes, growing more vegetables more often asks a lot of your land and soil, which is why I've spent the last couple of chapters discussing how to build up your soil's fertility. Use the techniques in this chapter, combined with what you've learned in the previous two chapters, and you'll be well on your way to producing more food for your family than you thought possible.

Crop Rotation

Planting different crops each year is known as *crop rotation*. When you plant different crops in the garden area each year, you are helping to keep your soil healthier, and you minimize the impact of pests and diseases. Wonder if it's worth the trouble to take an hour and jot out a rotation plan? Here are some things to consider:

- ✧ Soil health is improved when you rotate your crops because different crops take different types of nutrients and minerals. Rotating into the mix plants that actually feed the soil, like cover crops and legumes, will only increase the benefits!

- ✧ Crop rotating minimizes pests, especially for pests that overwinter in the ground or lay eggs in ground litter and debris.

- ✧ Disease infestations are greatly minimized when you rotate crops on a yearly basis. Fungi, molds, and diseases like root nematodes can build up their presence around a particular type of crop. By planting a crop from a different family the following year, these diseases don't have a chance to build up strength.

There are nine primary plant families that you'll want to understand when planning your crop rotation strategy. Each family has specific characteristics and are grouped as follows:

Nightshade family. Tomatoes, peppers, eggplants, and oddly enough, potatoes are in this family. These plants are all considered heavy feeders and will quickly deplete nutrients from the soil. Old-timers say these plants grow best following the corn family but I'm not sure I've ever heard the reason why.

***Brassica* family.** This family is also sometimes called the crucifers because of the small, four-petal flowers. In this family you'll find broccoli, cabbage, kale, mustard, and related plants like cauliflower and Brussels sprouts. All the plants in this family grow large leaves and require high levels of nitrogen to perform well, so a backyard farmer might plant these after the legume family.

Sunflower family. Light feeders, these plants can follow heavier feeders without much trouble. The sunflower or *Compositae* family includes lettuce, endive, and artichoke.

***Alliaceae* or onion family.** These include onions, leeks, garlic, chives, and shallots and are known to repel certain unwanted pests. While they can be heavy feeders, the bulbing plants can be kept well mulched with compost to help feed them. Use these to fill in any gaps in your rotation plan.

Pumpkin or cucumber family. As you can imagine, the cucumber family includes all the gourd-like plants such as pumpkin, summer squash, winter squash, melons, cucumbers, and gourds. Many of these plants take up a lot of room unless grown vertically on trellis structures. These are a heavier-feeding family and can go before the spinach family or carrot family.

Carrot family. These include carrots, celery, cilantro, parsley, dill, and fennel plants. These plants usually attract a large number of pollinators to the garden, benefiting the production of other crops.

Spinach family. This family is sometimes called the goosefoot family, and it includes beets, spinach, chard, and quinoa. These are lighter feeders and can follow one of the heavy-feeding families of plants.

Legume or pea family. These plants include peas, beans, lentils, and peanuts and are considered beneficial plants for the soil. Legumes help build up nitrogen in the soil and can be planted before heavy feeders in the crop rotation.

Corn or grass family. The grass family of plants includes corn, wheat, oats, and rye crops, although most backyard farmers are most familiar with growing corn. Tradition says to plant corn ahead of potatoes for the best potato harvest.

Crop rotation is relatively easy when you have a basic plan or guideline. The following two diagrams show how someone might rotate their growing spaces. You'll see that the plan is very customizable depending on what types of vegetables and herbs you grow, and how much you grow. Notice in the first figure that follows that our family has entire rows dedicated to tomatoes. We treat them like a family of their own in our rotation cycle.

Cucumbers, onions, tomatoes, beans, and potatoes in a small garden area the first year.
(Illustration by Becky Bayne)

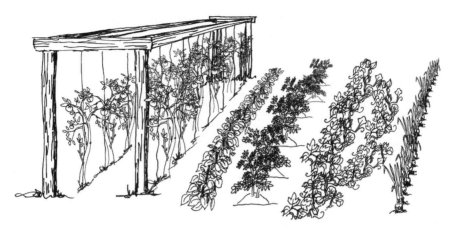

Each year you can step your crops over a bed or two (or a row or two in a smaller garden, like the example shown). This is what a garden might look like in the third year with tomatoes, beans, potatoes, cucumbers, and onions as each group of crops moved a row to the left each year.
(Illustration by Becky Bayne)

Your personal tastes will dictate exactly how your crop rotation plan looks, of course. Regardless of whatever plan you develop for yourself, here are some important points to keep in mind:

- ✤ Group plants according to their plant families for the easiest rotation plan. You can move your vegetables according to their plant family so that you are changing up the soil requirements, as well as the pests and diseases they are exposed to.

- ✤ Potatoes seem to be a unique type of crop. The country wisdom I've heard is that potatoes will produce more when they follow the sweet corn crop. Also realize that potatoes are in the nightshade family with tomatoes and eggplants, so don't plant them in ground where tomatoes were grown the year before.

- ✤ To increase self-sustainability of your land, consider having a portion of the land "lay fallow"—a season of rest. Many backyard farmers will plant a cover crop like clover that helps fix nitrogen in the soil. We like to plant a legume like peas or beans that have a similar effect.

Crop Succession

Succession planting is the other way to manipulate how your crops are planted to increase your harvest. Succession planting is the fancy term for it—I call it cramming as many plants as you can into a single year. When you have crops that grow best in different times of the year, or crops that grow rapidly, you can use that to your advantage by planting them one right after the other.

This can mean the same crop or another crop. For example, this year as soon as we harvested our broccoli, we were able to plant that same garden space with tomatoes. We also harvested our bush beans and immediately replanted with new bush bean seedlings we'd already started, allowing us to get a double harvest out of a single growing season.

ON A DIFFERENT SCALE

It is totally possible to practice succession planting on a very small scale. If all the space you have for growing produce is a large container garden, you can still pack it to the max. The same rules apply. Start with lettuce or other cool-weather greens and sow the seeds for basil indoors. After your first lettuce crop is spent, you can remove it from the container and quickly put in your basil seedlings. Successive planting works even in a tiny balcony or patio garden!

One of my favorite ways to plant successively is to plant a warm-season plant after a cool-season plant. Sometimes I see gardeners put their tomatoes out early in the spring, but tomatoes really do prefer lots of warm sunshine to grow their best. Start the tomato seeds indoors and let your cool-season lettuce, broccoli, and cabbage do their thing for a while.

After you've harvested your full cool-season crop and those plants rebel against the heat, you'll be able to take them out, lay down extra mulch and compost, and immediately replant with tomato, pepper, or eggplant starts.

The other way to plant in succession is to start seeds at different times. This allows you to always have a new wave of crops maturing, as you harvest the first set of plants. Vegetables that do well with this technique are usually fast-growing plants such as beans, radishes, and lettuce.

Both of these successive planting techniques used together can keep your garden healthy, vibrant, and always producing something for your backyard homestead.

Extending the Growing Season

One of the other ways you can produce more food on your backyard farm is to extend your growing season. The beginner version of this is planting seasonally—cool-season plants in the fall or early spring, and warm-season crops through the heat of summer.

But what if you're greedy like me and don't want to be tied to the typical growing season? That's when you get creative with a few more advanced tricks for extending the season. Let's look at several of these options.

Seasonal Plantings

Perhaps what comes to mind for you when you think about gardening is a typical summer garden—corn standing high, tomatoes ripening, and melons climbing trellises in the corner. However, that's only a few weeks of the entire year. There are many plants that tolerate, even thrive, in cooler weather and can be grown through the fall months as well.

These fall gardens can provide fresh produce through the entire year in some areas, and stretch the growing season an additional few weeks in others. The first key to success with a fall garden is to select plants that are suited to cool weather, such as the following:

Beets	Kale
Broccoli	Lettuces and arugula
Cabbage	Mustard
Chard and Swiss chard	Radishes
Green onions	Spinach

While you can purchase plant seedlings already started from the nurseries and garden centers, it is more difficult to do so for a fall garden. Many big box garden centers stop offering new vegetable seedlings around September or October, and in our area the only seedlings available were broccoli and cabbage. As you can see from the preceding list, two plants barely touch the beginning of what you can grow!

The best way to make sure you get the most variety from your cool-season garden is to start your own seeds. Some crops are only available through seeds most of the time. For example, the heirloom Purple Sprouting Broccoli is noted as being extremely winter hardy, if a bit slow-growing.

Cover Your Crops in the Garden

When the temperature drops suddenly, or you are trying to keep your tomatoes going through what is expected to be a temporary frost, you can cover the plants in the field. Use a thick layer of mulch, row covers, hooped row covers, cloches, or Wall O' Waters to provide a temporary layer of protection for tender fruits and vegetables. Each option provides varying levels of protection. Each also comes with pros and cons, so choose whichever option is the best for your situation.

Thick layer of mulch. Sometimes you can protect newly planted seedlings from a late frost, or fruiting crops from an early one, with a thick layer of loose mulch. It's easy to apply in a quick pinch, which is the upside. Of course, you'd need a supply of straw or mulched leaves at hand, and it won't work well if the weather is also very windy. Mulch can be expected to give just a couple degrees of additional protection, so it's best for a light frost situation.

Row covers. Row covers are sheets of fabric or plastic draped over the entire row of plants. They provide an insulating layer for the crops, and if the row covers are clear plastic sheets, they even allow sunlight to come through. Row covers make it easy to cover an entire row at one time, but generally need more than one person to apply to the garden. Plastic sheeting can build up heat inside if the sun comes back up the next day, so they'll have to be removed from the plants. Plastic sheets can also become heavy in rainy weather if water collects on top of them, causing stems of the plants to break.

Lightweight row covers can provide a minimum of 3 degrees of protection, but a University of Florida study showed that frost protection can be even higher. Heavier row covers can provide protection of 6 degrees or more. The best protection is provided to sections of the plants that aren't touching the cover, another benefit of using hooped row covers (discussed next) to elevate the cover.

Row covers that are laid directly on the ground can be secured with dirt, chunks of wood, or rocks.
(Photo courtesy of Broadfork Farm)

Hooped row covers. These are the same fabric or plastic sheets that can be draped directly over plants, except they are stretched over PVC hoops that hold the cover off the plants. These are more expensive, and a bit trickier to set up for the average small farmer. However, they don't have to be removed during the day, because the ends of the tunnels can be opened for ventilation. They also keep plants healthier because they don't bruise or break the plants by lying directly on top of them. The insulating sheeting can be replaced by shade cloth in the heat of summer as well, allowing the hoops to serve double-duty for backyard farmers.

Cloches. If you just have a few tender plants to cover, you can actually cover each plant individually. This is more labor intensive, but in a small garden that usually isn't a problem. Milk jugs with the bottoms removed can make excellent temporary cloches on the cheap (bury the edges into the dirt to keep them from flying away). I love the gorgeous, old-fashioned cloches made of glass that you can find sometimes. I've also used large glass pickle jars or mason jars with no labels on them. Just be sure to remove the cloche covers when the sun comes up so your seedlings don't overheat!

Wall O' Water. These are similar to cloches in that they individually surround each plant. Wall O' Water are applied individually around each plant and can provide protection down to 16°F to 20°F! Wall O' Water aren't closed at the top, however, so they give you a bit more leeway if you don't get them removed as soon as the sun comes up. They offer excellent protection from cold. The downside to these in the backyard farm is that they can be costly to purchase, and labor intensive as each unit has to be filled with water and placed. On a larger scale, this is probably not the best option.

Cold Frames

A more permanent and effective step up from a row cover is a cold frame. A cold frame is a box, topped with glass, which is set into a garden space. Plants are then grown inside, either planted directly in the ground, or grown in containers housed inside the cold frame.

ON A DIFFERENT SCALE

When I say that cold frames can be amazingly simple, I mean it. One year, I repurposed an old fish tank that had cracked but not broken as a temporary cold frame structure when a late frost caught me off guard. I simply flipped the aquarium upside down over my newly sprouted seedlings and didn't lose a single one! Not fancy, by any means, but it worked in a pinch.

Cold frames can be very elaborate, or amazingly simple. At the most costly and permanent end of the spectrum, I've seen bricked boxes with hinged windows on automatic openers. At the simplest end of the spectrum I've seen hay bales arranged loosely into a square shape with plastic stretched across the top. Both are surprisingly effective means for extending your growing season.

This cold frame is angled to allow more sunlight to shine in. You would want to face the low side toward the south if gardening in the Northern Hemisphere.

(Photo courtesy of Steve Laurin)

Notice how this cold frame has hinged glass panes that can be opened for better ventilation. More simple boxes have to be opened by lifting the glass pane off the top.
(Photo courtesy of Steve Laurin)

Make sure you allow ventilation in your cold frame when the sun comes out, or your poor cold-hardy plants will bake to death. I'm not saying I've done this—just giving you a friendly warning.

See Appendix A for a simple cold frame plan you can build yourself to save money.

Greenhouses

Greenhouses are a year-round way to help regulate the temperature of your plants so you can extend your growing season. Some are even heated or lighted in the coldest winter days to allow year-round production. I have friends living in mild zones who have been able to produce tomatoes throughout the year. At the very least you'll be able to add weeks to your growing season!

Greenhouses vary even more than cold frames in their design and expense. The most elaborate are huge walk-in structures with heat, automated ventilation systems, and large walkways. The most simple are barely bookshelves covered in plastic and don't allow the gardener room to come inside at all. And of course, there is everything in between. As with all garden accessories and structures, you can buy kits for a higher price, or build your own as a way to save money.

If you're handy, you can build a simple but effective greenhouse rather easily. A wooden shed structure about 4×8 can be built for just a couple hundred dollars. The frame can be wood with greenhouse plastic stretched around it to allow the sunlight through. Of course, prebuilt options are available commercially in almost any size and design imaginable.

This large greenhouse is the centerpiece of the year-round garden at the Summer Winter Restaurant in Burlington, Massachusetts. With it, Rachel Kaplan, an in-house gardener, provides a variety of fresh produce.
(Photo courtesy of Summer Winter Restaurant)

Start Plants Indoors

Another way to extend your growing season is to start your seeds indoors. By getting the jump on the season in this way, you'll have seedlings ready to go in the ground as soon as the soil is warm enough. The obvious thing is to start seeds in late winter or early spring for your spring and summer crops. But don't forget that you can start seeds indoors during the late summer as well. Then you'll have them ready to put in the ground for your fall garden!

Starting seeds indoors allows you to plant in succession as well. Start a new round of fast-growing plants like lettuce and beans every two to four weeks. That way, you'll have new seedlings ready to go!

Companion Planting and Interplanting

Have you ever seen old pictures of cottage kitchen-gardens that have herbs mixed in with their vegetables? There are benefits to growing certain plants together instead of a single crop over the entire space. *Monoculture* creates the perfect environment for outbreaks of pests and diseases, while *companion planting* allows plants to work better together than they would individually.

DEFINITION

Monoculture is when a farm area grows only a single crop. They are more susceptible to soil deficiencies and pest problems. **Companion planting** is purposefully planting one type of plant beside another to increase the flavor, growth, or protection from pest and disease. **Interplanting** is planting one type of plant within the same space as another. It's more than planting one type of plant in the row next to another—plants actually share the row or garden plot.

Planting beneficial herbs near your vegetables can help attract pollinators and discourage pests. Many of the plants that are often used as companion plants are flowering herbs such as nasturtium, marigolds, and basil. The hardest part is sifting through the lore, magic, country tales, and what has been born out by research and study.

French marigolds (*Tagetes patula*) are one of the most commonly cited plants to include in companion plantings for a variety of reasons. They were shown in a study by the University of Georgia to help decrease the amount of root-lesion nematodes in the soil. They also attract pollinators to the garden and have been shown to repel whiteflies.

Tansy attracts ladybug and lacewing predators to the garden where they happily munch on aphids. Aphids are a common pest of tomatoes and roses especially. Tansy was also shown to decrease the number of cucumber beetles by more than 60 percent in a study by Rodale Institute Research Center. Cucumber beetles prey on all members of the squash family, and also on tomatoes and potatoes.

European corn borer is a common pest of pepper plants, corn, beans, and tomatoes. A study conducted at Snyder Research and Extension Farm over the course of two years showed that dill, coriander, and buckwheat all had positive effects in decreasing the amount of fruit damage from European corn borer.

Another form of companion planting is known as *interplanting*. The iconic example of interplanting is known as the three sisters, and was perfected by Native Americans. Corn, a shallow-rooted plant, was interplanted with beans, which are medium-rooted, and squash, which is deep-rooted. The beans could climb the stalks of the corn as support, while the squash provided many of the benefits of mulch with its broad leaves.

The trick when interplanting is to consider the timing. If you plant carrots and radishes together, for example, you would have time to harvest the fast-growing radishes before the carrots are too old. However, if you tried to pair a slower-maturing crop like tomatoes with carrots, the tomatoes would shade out the carrots and inhibit their growth.

One plant that is commonly used in interplanting situations is white clover. White clover is a nitrogen producer that doesn't grow too tall to compete for sunlight. It is also used as a living mulch for crops that consume a lot of nitrogen, such as tomatoes, corn, broccoli, spinach, and cabbage.

These beans (to the left) are interplanted with lettuce (in the background). When the lettuce finishes in the spring, the beans will be ready to grow all summer. Cilantro (to the right) grows at the end of the row and will benefit from staying shaded by the beans in the heat of summer.

Getting the most from your garden means planting in smart ways and respecting the land. Now let's get on with the fun stuff—the plants!

Vegetables for the Backyard Farm ~ 8

Now that you've created a healthy foundation for your garden (Chapter 6), and you're equipped with a plan to make the most of every inch (Chapter 7), we're going to look at some of the most popular vegetables. While most people think of vegetables as the main part of a seasonal garden, experienced homesteaders will tell you there's a lot more to a diet, and thus the backyard farm, than just a handful of veggies.

Vegetables to Grow on an Acre

There is something miraculous about growing a garden. A single seed takes dirt, sunlight, and water and transforms itself into an exponential harvest. A truly compounding investment, gardening can bring a large percentage of the meal's food straight from the backyard farm. This section digs deeper into the most popular, and most valuable, vegetables for your garden.

We not only look at which vegetables to grow, but also why, what makes them special, and when you should think about adding one or more of these to your garden. In each section I also discuss some specific varieties to consider adding to your kitchen table.

The vegetables mentioned here are only an introduction to the many varieties that are available. Be adventurous with your garden!

Cool-Season Vegetables

These vegetables grow in the cooler seasons and will often bolt, or go to seed, during warmer weather. They can be grown in early spring, or often are planted in late summer to grow for a fall crop. Growing them in a fall garden is especially useful in places where a mild spring doesn't last very long before summer heat turns brutal.

Beets (*Beta vulgaris*)

Beets are a root crop that is easy to grow and produces well. Beets are useful as food for not only the humans on the homestead, but as feed for the animals also. The sweet roots of sugar beets are used as a source of sucrose sweetener, and with good reason. Beets are a crop that is often overlooked in many diets, but has the potential to be a main staple in the homestead kitchen. Try growing cultivars that are white, yellow, or orange if you don't like the red color that can stain hands.

Beet seeds are actually the fruit of the plant and are multigerm, which means more than one plant will sprout from a single seed. Sow one seed at a time, and space them about 3 inches apart. In the spring, you can begin planting 3 to 5 weeks before your last frost date, and in the fall start your last round of planting about 10 weeks before your first frost date.

When the seedlings emerge, you will notice that many of them have three or more seedlings in one clump. Thin out the weakest ones so that each clump has one strong seedling growing. Be sure to toss your thinnings to the chickens or goats or at the very least into the compost bin so you aren't wasting them.

After the beets are well-started and have laves about 4 inches tall, go through and thin the survivors 5 inches apart. Beets will continue to grow as long as they have space and friable soil. Beets will be ready to harvest anywhere from 7 to 13 weeks or more depending on when in the season you're growing and how the particular variety grows.

Beets are a drought-tolerant crop with enormous roots. Give them room to spread and they will do well even in poor soil. In fact, if the soil is too rich in nitrogen, your beets will produce large tops but broken roots. Beet tops are sometimes used as greens in the kitchen but I prefer chard or kale, and let the livestock eat the beet tops instead.

Harvest beets by selecting the largest ones first and harvesting as needed. This will allow more room for the smaller beets to continue growing. I know some farmers who simply dig every other one to get the maximum growing space for their beets.

OVER THE GARDEN FENCE

Beet tops and roots are often used as livestock fodder in small-scale productions trying to avoid using commercial feed. Beet tops can be twisted off the roots, dried, and then fed to any livestock on the farm. Whole beets should be broken, ground, or cut up in some way to avoid cattle and pigs choking on small beets. I've heard of ranchers using wood chippers, feed grinders, and even driving over beets to crush them before feeding to livestock.

The roots you harvest in the fall will store for several weeks in a cool root cellar. Roots harvested earlier in the season can be eaten fresh or even pickled and canned. And don't ignore the tops, which can be eaten as greens similar to kale or collards!

Beets tend to resist most pests in the garden. Digging rodents can steal your crop sometimes, and the edible green tops are a lure for deer or wild rabbits. Leaf miners sometimes attack beet leaves but are not considered a huge problem in most areas.

Albino beet. As the name implies, the albino beet has a white-fleshed root.

Bull's blood. A deep red beet with pinkish rings inside, the leaves are an attractive reddish-purple color. About 10 weeks to mature.

Chioggia. This gorgeous Italian heirloom has red- and white-striped flesh.

Heirloom Chioggia beets have beautifully colored roots.
(Photo courtesy of Baker Creek Seeds)

Giant Yellow Eckendorf. A huge rooting beet, the flesh is a golden yellow. The beets grow to be several pounds.

Ruby Queen. A popular canning variety that performs in poor soil and tolerates intensive planting.

Broccoli (*Brassica oleracea*)

Part of the *Brassica* family, broccoli is a vegetable high in vitamin C and calcium, which makes it a great choice for home gardens. A fast grower, broccoli prefers cool weather as it will not tolerate as much heat as the similar-looking cauliflower will. Broccoli grows easily from seed and can be planted in midsummer through fall, following the frost-tender vegetables. I love broccoli for extending the total produce grown in the garden each year.

Broccoli grows from seed to harvest in about 11 to 14 weeks although plants started in the fall will grow a bit slower than plants started early in the spring. They don't like having their roots disturbed so start seeds in plantable pots. Begin hardening off the plants to move to the garden at about four weeks.

If you're interplanting with summer crops, cut those vines/plants off at ground level after they are dead instead of pulling them up by the roots, so you don't disturb the root system of your broccoli plants. In this case you can plant the seeds directly in the soil about ¾ to 1 inch deep. I allow about 14 to 20 inches between my broccoli plants, which is pretty close spacing, and this allows the broad leaves to shade out weeds beneath them.

Grow your broccoli in soil that is between 6.0 and 7.0 pH and make sure they get full sun. Broccoli will take plenty of nutrients so rotate it in after a lighter feeder like onions or beans, and be sure you've applied plenty of compost or organic fertilizer (see Chapter 6). This same tendency to feed heavily will make your broccoli sensitive to weeds, so be sure you mulch thickly enough to smother out the weeds in your garden, and pull any stragglers in your garden area once a week.

All broccoli should be harvested before the green florets bloom into yellow flowers. When the yellow begins to show it is really too late, and it can happen almost overnight in the spring so watch out!

OVER THE GARDEN FENCE

I learned the first year we grew our garden that fresh broccoli is more tender than what I was used to buying in the grocery store. When you cook your garden-grown broccoli, you won't need to cook it as long because it is fresh and harvested when ripe!

If you're harvesting heading broccoli you'll get one main, large head, so let it grow big enough to provide a decent harvest. Sprouting broccoli will produce side-shoots of broccoli as you harvest mature heads, so the more you harvest, the more it produces. Broccoli raab (pronounced *rob*) is usually harvested for its zesty greens and should be harvested before the florets mature to avoid a bitter taste—usually within a mere 50 days of planting. Romanesco varieties have unique spiraled heads and generally require a little more growing room.

Pests that attack broccoli are usually the same pests you'll find on cabbage. Army worms, flea beetles, cabbage worms, and cabbage aphids can all attack the leaves. Cabbage root fly is a pest that can attack the root of the broccoli plant causing a sudden decline in a plant that otherwise seemed healthy. Rotating where you plant your cabbage family plants can help break the lifecycle of this pest and prevent attack.

Di Ciccio. An Italian heritage variety that offers heavy production of side shoots after the head has been harvested.

Early Purple Sprouting. This broccoli is a sprouting type with lots of side shoots. The broccoli heads are purple instead of green.

Small Miracle. One of the most compact broccoli plants available, this is one of the cultivars you could grow in a container.

Waltham 29. Pictured below in my garden, this broccoli produces side shoots in addition to the main head.

Most broccoli varieties produce a large head, but many old-fashioned varieties will produce additional side shoots.

Cabbage (*Brassica oleracea*)

Cabbage is a bit touchier than broccoli and will perform much better when planted for the fall growing season. Our Oklahoma springs generally don't last long enough for our cabbage to form heads so I no longer plant it at the beginning of the season. Cabbage is one of the most cold-hardy vegetables and tastes better after a touch of frost.

Cabbage is a perfect example for why building your soil's fertility naturally is so much more effective—when cabbage is fertilized too heavily and too quickly it can cause cracked heads that are weaker. Fertility built into the soil will be released as the plant needs it compared to the synthetic fertilizers that are poured over the top of the plant.

Cabbage can be started indoors and transplanted to the garden a few weeks before the last frost date. Seeds only need to be about ¼ to ½ inch deep and planted 12 to 24 inches apart depending on whether they are an early maturing variety or a late-maturing variety. Cabbage doesn't develop deep roots so be sure you water consistently to avoid stressing them.

Cabbage adds color to the fall garden as it's available in both green and red varieties. Late-maturing varieties are much larger and can grow as big as a bowling ball at 6 to 8 pounds! Early maturing varieties are smaller and can be harvested much sooner.

Brunswick. Large, drumhead cabbage that is historically grown both for market and for long-term winter storage.

Glory of Enkhuizen. An heirloom variety known as a good sauerkraut cabbage. Medium-large, bright green heads mature in about three months.

Mammoth Red Rock. Deep red cabbage with huge heads are both colorful and flavorful.

Perfection Drumhead Savoy. Slower to mature, this large drumhead has crinkly or *savoyed* leaves. The leaves are milder and after a touch of frost, become very sweet.

DEFINITION

When talking about plants, the **savoyed** leaves are those that are crinkled. Sometimes you'll see the term "semi-savoyed," which simply means "a little crinkled."

Savoy di Verona is an example of the variety of cabbages available if you use heirloom varieties and grow them from seed. This cabbage is variegated green and red with a pleasing mix of colors.
(Photo courtesy of Baker Creek Seeds)

Lettuce (*Lactuca sativa*)

One of the most pleasing surprises to me when we started our gardening journey was the amazing variety of lettuce available. Colorful! Delicious! Zesty! Sweet! The world of Iceberg lettuce is like black and white television—you don't know what you're missing until you switch to high-def color. There are four main types of lettuce you can try in your garden:

- *Crisphead lettuce* is the type of lettuce you tend to think of at the grocery store like Iceberg. These are the most heat tolerant of all the lettuces but will also tolerate cool fall weather.

- *Butterhead lettuce* has hearts as well as loose leaves. This type is drought and heat tolerant compared to many of the other varieties.

- *Romaine or cos lettuce* has an upright form and doesn't have a thick heart like the previous two types. Hardy in cold weather, the romaines generally do not tolerate heat well.

- *Loose-leaf or salad bowl lettuce* includes a huge variety, such as black-seeded Simpson, Oak Leaf, and Red Sails. As the name implies, this type has loose leaves instead of a heart to the plant. It is the least heat tolerant.

Heirloom lettuce, like this Devil Ears lettuce, provides a myriad of colors and tastes not available in the grocery store.
(Photo courtesy of Baker Creek Seeds)

Lettuce can be grown anywhere that gets enough water to support it because periods of drought will cause bitterness in the leaves. When sowing seeds outside you want the soil temperature between 40°F and 75°F, because if the weather is too warm the seeds won't germinate. Late fall or early spring is ideal for most of the lettuce types, especially in the warm southern climates.

Sow seeds indoors when outdoor temperatures are not quite ideal and then transplant to the garden at four weeks. Plant lettuce every couple weeks so you'll have a plentiful harvest as the season progresses. I like to try to plant lettuce about 6 to 12 inches apart so I won't have to thin it very much. The key is to plant a lot of different varieties of lettuce because there are so many colors and tastes to choose from.

OVER THE GARDEN FENCE

The darker the leaves of your lettuce (whether red or green) the higher in vitamin A that variety will be.

Lettuce can be harvested from the outside leaves in (except heading lettuces), so don't be afraid to begin harvesting as soon as you get hungry but leave the growing center. Lettuce that has a heart should be harvested whenever the heart forms so that the lettuce won't bolt, or go to seed.

Lettuce has a lot of pests that will attack the leaves. Aphids are one type of pest in particular and can be controlled by allowing predators to thrive in the garden. You can also rinse them away with soapy water spray. Slugs can be a huge problem with lettuce, and this is one case where a thick layer of mulch right around the base of the plant might be a problem and not a help. If you see chewed leaves and slimy trails you'll know what your problem is. Beer traps or upside-down orange halves set nearby will help you naturally control the slugs in your garden if you're prone to them.

Amish Deer-Tongue. An elongated leaf, and loose habit, has made this slow-growing lettuce a favorite in backyard farms for generations.

Black-Seeded Simpson. One of the most popular heirloom lettuce varieties, it matures early after only 45 days.

Buttercrunch Bibb. A popular heirloom variety, this cultivar was named All-American Selection when it was introduced in 1963.

Flame. A strongly colored red lettuce that is becoming more popular as a local market and gourmet offering.

Gotte Jaune D'Or. Also called Golden Tennis Ball, this loose-leaf lettuce is a bright, lime green.

Little Gem. A butterhead-type lettuce with small, compact heads growing only about 6 inches tall.

Oak Leaf. Uniquely cut leaf shape resembles an oak leaf, giving this slow-to-bolt variety its name.

Paris Island. A sweet-tasting Romaine type of lettuce with dark green outer leaves and a pale-green heart. Great for year-round planting.

Rouge D'Hiver or Red Winter. Highly cold-tolerant lettuce from the early 1800s, the leaves are dark purple-red.

Ruby Leaf. This is a highly attractive leaf lettuce with variegated red leaves and a green heart.

Mustard (*Brassica juncea*)

India mustard is grown for the delicious and nutritious leaves. I love adding chopped mustard to our stews in the winter to help boost the overall vitamins we are consuming during the times we have fewer fresh fruits and vegetables coming from the garden. High in vitamins K and A, mustard is also high in calcium and iron.

From seed-sowing to maximum harvest is about 12 weeks but you can begin cutting leaves within just a few weeks. Seeds will not sprout when temperatures drop below 45°F and should be planted ¼ to ½ inch deep, much like lettuce. Plant mustard about 12 inches apart and you'll find this plant a carefree, easy-to-grow addition to your fall garden.

As with lettuce, drought will cause the mustard plants to turn bitter and sharp. Harvest leaves around 45 days of maturity so you'll have tender greens. Mustard can be eaten raw, which will have a more peppery flavor, or steamed lightly to bring out a natural sweetness.

Mustard has fewer pests than lettuce or cabbage, even though it's in the broccoli family, too. If flea beetles become a problem, try adding a row cover. Club root can develop if you plant mustard in the same area year after year—so get out that garden journal and rotate your crops (see Chapter 4)!

Japanese Giant Red. This variety has purple-red leaves with green centers, and a strong, sharp flavor. Good for stir-fries.

Japanese giant mustard, an heirloom variety, has multicolor leaves.
(Photo courtesy of Baker Creek Seeds)

Southern Giant Curled. This heat-tolerant, American mustard has curled and savoyed leaves that grow upright. Foliage is bright-green and attractive.

Tendergreen. A quick-maturing mustard, the tendergreen has deep-green, glossy-looking leaves.

Peas (*Pisum sativum*)

Peas are a great snack fresh from the garden and provide high levels of vitamin C and iron. Low-maintenance in the garden, peas don't usually need extra fertilizer. Peas stop producing in the highest temperatures, so in areas with hot summers you'll want to plant them in spring and later summer.

We started out planting peas in double rows on either side of a trellis and allowed all the plants to grow up. I read in *Mother Earth News* a suggestion to plant a shorter-growing variety beside the longer, rambling pea vines. Then the quick-growing short vines mature more quickly as well as helping to support the longer variety pea vines as they grow.

Sow your peas about 1 inch deep in soil that is not too cold or wet or the peas will rot. Spring plantings are often better started indoors and then transplanted outdoors when they have a good start of a couple inches. In the late fall, plant outdoors where the warm soil will help with germination. Peas can be soaked overnight to help increase the germination rate as well.

Peas mature between 11 to 14 weeks depending on the variety. There are four main types of peas. Snap peas are eaten whole with the peas still in the pods and are harvested when the peas just barely begin to show. Soup peas, on the other hand, are grown for drying and storing and so are harvested when they have dried completely on the vine. Snow peas produce tender pods, eaten whole, and are often added to stir-fries and salads. Shell peas, or English peas, are grown until the pea pods are full and are shelled from the pods.

HARVESTING PEAS

Use care when harvesting climbing peas so you don't break the clasping tendrils when you pull off the pods. Hold the plant steady with one hand and pluck the pods off the vine with the other. Pick peas often to encourage healthy production and preserve extra harvest by freezing, canning, or drying.

Peas are sometimes eaten off the vine by birds and rodents. Mosaic virus can also be a problem in some regions, especially where peas are overcrowded or lack good airflow. Root rot is a problem where fungus gathers in the soil, and rotating crops can help minimize that problem.

Alaska. One of the earliest maturing peas, it isn't always as tender as some of the other peas, but makes a great soup pea. Fifty days to maturity makes for a short growing time.

Lincoln. Also called the Homesteader, this pea is highly heat tolerant and wilt resistant. Compact vines produce loads of easy-to-shell peas.

Mammoth Melting. A sweet snap pea with wilt-resistant vines up to 6 feet long. Best in cool weather.

Sugar Ann. This sweet, edible pod pea took the All-American Selection prize in 1984 and grows on a more compact, container-friendly vine of 4 feet or less.

Tall Telephone. Also called Alderman. This prolific vine was named for Alexander Graham Bell and is great for canning and freezing.

Wando. A versatile pea that will pollinate well in cool weather, and holds up better in hot weather than most peas. My kids call this the Waldo vine.

Spinach (*Spinacea oleracea*)

There are two main types of spinach to try in your homestead garden. The first group of spinach is savoyed or semi-savoyed, which is the type of spinach with crinkly leaves that are a very dark green. Smooth-leaved spinach is a lighter green and has smooth leaves instead of the textured leaves.

In southern gardens, you can sow seeds in the fall about five to seven weeks before the first frost date, and you can begin harvesting as baby greens very quickly. In northern gardens, spinach will behave well when planted in the spring and not bolt too soon. Either way, plant your seeds ½ inch deep and 8 inches apart in loose, fertile soil. Spinach is one of the cool-season plants that is not in the *Brassica* family, so you can plant it where tomatoes grew in the summer or where broccoli and cabbage were the previous year.

You can begin harvesting delicious leaves from the outside of the plant as baby greens within just a couple weeks (as soon as plants reach 2 inches tall). Continue harvesting as needed from the outside of the plant until the spinach begins to bolt, or cut the whole plant when it reaches 6 to 8 inches tall. Sow new seeds every couple of weeks so you'll have a continuous harvest through the fall and well into winter in some areas. Unlike lettuce, spinach freezes well so be sure to plant enough during the prime fall season to last through the summer when spinach won't grow well—or in my area, at all.

Spinach doesn't have very many pests but leaf miners can be a problem sometimes. If you see pale tunnels in your leaves, remove infected leaves and cover the plants with a floating row cover to keep the flies off the spinach. Slugs can be a pest on spinach, too, and if you see the leaves being chewed from the outside, try setting beer traps to control the overall population.

Bloomsdale. A long-standing spinach with excellent heat resistance, this spinach was first developed in the late 1800s.

Giant Noble. Huge spinach grows with a loose, open habit and outer leaves can be harvested in just a couple weeks.

Monstrueux De Viroflay. This spinach develops huge leaves nearly a foot long and is designed to grow quickly in the fall before winter weather sets in full force.

Swiss Chard (*Beta vulgaris*)

Swiss chard is in the same family as spinach and beets. In fact, Swiss chard is basically a beet that is grown for its stem and leaves instead of the root. And wow! What a colorful and tasty stem and leaves it produces! Swiss chard is one of those plants that is as much at home in the front planter bed as it is in the vegetable garden, because the brilliant colors are so attractive.

Sow your seeds outdoors when the soil is at least 47°F or sow indoors and transplant out in early spring. Plant seeds to 12 inches apart in soil mixed with new compost. Some swear by soaking the seeds overnight to help improve germination, so that's something you can try if you want. Seeds will last for three years when stored properly, so don't feel like you have to plant the whole packet—you can plant half and save the rest for the following year.

Swiss chard produces for a long time in the fall garden. You can harvest the outer leaves only when the plants are newer. On plants that are more mature, you can harvest the entire plant by cutting off stems about an inch above ground level. Swiss chard often regrows with new, tender stems that you can continue to harvest.

Keep Swiss chard watered consistently and mulch well, both to preserve the moisture and to keep water from getting on the leaves. If your growing season gets extended because of earlier planting in the summer or a mild early winter, renew the soil with an application of fertilizer and compost.

Swiss chard can be used as both a green and an asparagus or celery replacement in the kitchen. Stems can be lightly steamed, sautéed, or pickled, while greens can be dehydrated, blanched and frozen, or used in stews, omelets, and casseroles.

Leaf miners can be a problem for chard, as with spinach, and floating row covers are an easy way to help prevent this problem. If your seedlings aren't growing well it could be because of weeds in the area, because Swiss chard doesn't compete well with weeds. Clear your planting area before sowing seeds or transplanting, and then cover with a layer of mulch to smother out any weeds.

Bright Lights. Also called Five Color Sweetbeet or Rainbow Chard, this cultivar is quickly growing in popularity as a specialty market chard. Stems are produced in red, pink, yellow, orange, and white!

Flamingo Pink. This chard's name is not an exaggeration; the stems are bright pink, almost neon colored.

Fordhook Giant. Burpee, a plant and seed company, introduced this heavy yielding choice with dark green leaves and white stems.

Ruby Red. My kids call this Christmas chard because the stems are cherry red while the leaves are dark green.

Warm-Season Vegetables

Warm-season vegetables enjoy basking in long days of sunshine. The longer days and warmer soil temperatures are what these plants need to produce the classic garden harvests of tomato, corn, and zucchini.

Beans (*Phaseolus vulgaris*)

Beans are a fabulous plant to add to the garden because as a legume they add nitrogen back into the soil. Add beans to your crop rotation in between heavy feeders such as tomatoes and broccoli. Harvest beans early for fresh green bean produce, or allow them to dry for long-term storage. Some beans have a vining growth habit while others have a nonvining, bushy growth habit.

Beans are high in protein and in many cultures they make up a large part of the daily diet. On the homestead, a stored supply of beans can make it easier to fill in the gap during times when the freezer is a little lower on meat. Even with literally hundreds of varieties of beans, growing them is relatively simple and easy to manage regardless of which type of beans you want. Many varieties can even be harvested as green or snap beans early in the season, and then allowed to dry for a soup bean harvest as well.

Beans are easy to plant directly in the garden when soil is at least 55 degrees. Plant the beans about 2 inches deep. Space them about 5 to 12 inches apart depending on the cultivar and whether you're growing bush varieties or using a trellis for pole beans. Some more aggressive varieties may need to be spaced even farther apart, but with vertical garden techniques you can get away with more intensive planting.

Beans have a variable time to maturity depending on what variety you've planted, but generally you can begin picking snap beans within 8 to 12 weeks. If you are harvesting dried beans, allow around 12 to 16 weeks until the beans have dried in the pod on the vine.

OVER THE GARDEN FENCE

You can really maximize your snap bean harvest by picking your beans on a daily basis. The more frequently you pick the fresh beans, the more your plants will produce. When even a few of the bean pods reach maturity, your bean plants will stop flowering and cease production. Beans are commonly planted in succession so you can grow a new crop every couple weeks.

Dwarf bush beans will produce beans early, while pole beans will produce later in the season. Plant a variety so you can harvest throughout the entire summer. Beans are easy to can, freeze, and store dried.

Mexican bean beetles cause lacey patterns in the leaves where they feed and in severe infestations can even kill plants. Because the beetles overwinter in leafy debris, a tidy garden space with leaf litter, well-composted in a compost bin, can help break up the lifecycle. Parasitic wasps prey on these pests. Aphids and red spider mites can cause problems for bean plants as well. In both cases insecticidal soap sprayed on the leaves can help prevent infestation.

Contender bush bean. One of my favorite green bean varieties, this bush bean is prolific and produces early beans for fresh eating.

Jacob's Cattle bean. This colorful bean is a bush bean that is white and brick red in mottled splashes. The beans mature to fully dried beans in 100 days.

Montezuma Red. Also called Mexican Red, this dried bean bush is a sprawling vine with fabulous production. Beans are deep red and tasty.

Purple-Podded beans. In my region, these are called purple-hulled peas and the variety we have in our family is a bush form. There is also a pole form that grows large vines. Both are highly prolific and good as early snap peas, shelled beans, and dried beans.

These purple-podded pole beans are easy to grow vertically. My in-laws often grow a purple-hulled bush variety at the ranch to feed both humans and cattle.
(Photo courtesy of Baker Creek Seeds)

Tongue of Fire. An Italian heirloom with pinto-type beans is a perfect drying bean. Flavorful when eaten fresh or as dried beans.

Topnotch. This bush bean produces tender, yellow pods that are perfect for eating fresh off the vine.

Carrots (*Daucua carota*)

Carrots are a biennial root crop. The plant's natural cycle is to develop a large, thick root the first year and then move quickly to flowering and seed the following spring. Carrots are related to the wild Queen Anne's lace plants but bred for sweet, tender roots.

Carrots need loose, friable soil but grow better roots when they aren't fertilized too heavily with nitrogen. Heavy nitrogen encourages thick top growth and too-fast root growth, which can lead to forking and splitting.

Carrots do not transplant well, so plan to sow seeds directly into the garden. These are one of the few plants I really do sow thicker than needed because my carrots never seem to germinate as well as the other plants I like, and have a higher die off in the first couple weeks as seedlings. Thin gradually so that at a month you have seedlings 4 to 6 inches apart.

THINNING SEEDLINGS

Some crops can be thinned by pulling the plant out and tossing to the chickens, goats, or into the compost bin. Others, like carrots, should be thinned by cutting off the plant at the ground level. We use a simple pocket-knife for this and just slice it through the soil severing the stem from the roots. The idea is to not disturb the roots of the surrounding plants while they are growing.

For the best, most tender carrots you want your plants to grow quickly and easily. If the roots have to work too hard to get through the soil, they will be tough and fibrous. If your soil is more clay, consider the shorter carrot varieties.

Anywhere from 9 to 20 weeks will be needed before harvest, depending on the variety and how soon you harvest. Pull carrots as needed for fresh eating, or at the end of fall for root cellar storage. The carrot roots can be stored in a basket of moist sand.

Atomic Red. Brilliant red carrots are high in beta-carotene and lycopene and grow about 8 inches long in 70 days.

Chantenay Red Core. This variety tolerates heavier soil than most carrots and is a super-sweet, 5-inch-long carrot.

Danvers Half Long. A tapered wedge-shaped carrot that grows 6 inches long. This carrot has been popular for more than 100 years. Good winter-storage and very tolerant of a variety of soils.

Parisienne. A French, round carrot grown for market, it's a good choice for heavier soils.

Snow White. A white-fleshed carrot that matures at about 8 inches with a sweet, crunchy texture.

OVER THE GARDEN FENCE

In southern areas with mild winters, you can leave the carrots in the ground under a thick layer of mulch instead of harvesting them in the fall. Dig and harvest through the winter as needed!

Corn (*Zea mays*)

Corn is actually a grass on steroids. It's an improved grain and there are a couple different types. Field or dent corn is often grown for grinding to use as corn meals; sweet corn is what you usually think of as summer eating corn; and popcorn is a relative of the field corn and fabulous for snacking. Corn needs lots of water so its shallow roots don't dry out.

Sow your seeds directly into the garden and watch out for crows and other hungry birds. We've seen them go down the row pulling just-sprouted seedlings up by the leaves to get to the seed below. Sow seeds at least an inch deep when the soil is warm because late frosts will kill the leaves. Plant corn in wider rows to help with pollination, because if you don't have enough corn the fruit won't form as well. Space plants at least 18 inches apart.

Corn is shallow rooted and feeds heavily so it benefits from added compost throughout the season. The extra soil built up around the corn stalk helps stabilize the corn plant and the added nutrients help feed it. This means corn also doesn't compete well with weeds, so mulch deeply to help smother out weeds. In windy regions be sure to plant your corn rows where they will have some shelter from prevailing winds.

Corn is pollinated by wind, which is why planting in blocks is so helpful. You can also shake the corn stalks to loosen pollen and increase pollination. When the corn silk turns brown at the end, it's a good sign that the corn is ripe.

Sweet corn begins to turn starchy as soon as it is picked, so for fresh eating there is nothing better than garden-to-table in a single day. Super-sweets will also not produce as sweet as possible if they cross-pollinate with anything other than super-sweet corn. Allow plenty of room between the different varieties if you grow more than one kind of corn.

Corn pests include large animals such as mice, crows, and raccoons. There are also plenty of insect pests such as earworms and corn borers, which will eat their way into the corn and destroy the harvest. *Bacillus thuringiensis* var. *kurstaki* or BTK can be applied to the leaves or tips of the ears where these pests are a big problem. Gardeners who are trying to avoid chemicals can use a drop of mineral oil on the silk to stop corn borers.

Blue Hopi. An ancient blue corn that was used for flour and corn meal. Matures in 110 days with 9-inch-long ears.

Country Gentleman. A sweet corn with pale yellow, 8-inch-long ears. Good for fresh eating and also freezes and cans well. Reputed to have high germination and pollination rates.

Golden Bantam. An early maturing corn that produces yellow, sweet corn in 75 days. The bantam corn only grows about 6 inches long on 6-foot stalks.

Painted Hill Sweet. Developed for cooler weather, this corn matures in 75 days and has multicolored kernels of red, blue, yellow, and more.

Cucumbers (*Cucumis sativus*)

A favorite at our house, cucumbers are grown for eating fresh from the garden as well as for canning and pickling. Cucumbers are high in vitamin C and have a high water content, making them a classically refreshing treat from the garden. Think outside the traditional pickling cucumber when you choose varieties for the garden and look at cucumbers such as Asian cucumbers, which are beetle resistant.

Traditional gardening articles suggest growing cucumbers on hills and letting the vines ramble, but many gardeners find better success trellising cucumbers. Sow seeds directly in the garden after the other squash and melon seeds have already sprouted, when the soil is nearly 70°F. Plant seeds ½ inch deep and space them about 2 feet apart when growing up a trellis.

Begin harvesting cucumbers when they are large enough and don't let them reach full maturity. If the seeds inside one of the cucumbers fully mature, the plant will stop producing new fruit. However, if you harvest cucumbers from the vines on a daily basis, you'll get the most from each plant.

THORNY MATTERS

Stem rot and mosaic virus can be a problem if your cucumber vines are planted too early in the damp spring or if the vines are overcrowded. Growing up a trellis and allowing enough airflow around the plants can help prevent this. Cucumber beetles are often the biggest pest problem for cucumbers in the garden. Parasitic nematodes in the soil can help control the larvae and break the feeding cycle.

Lemon Cuke. A very unusual variety that grows round and yellow, resembling a lemon. Lemon Cuke has a mild flavor.

Marketmore. Matures in about 70 days; this cucumber has excellent yields and flavor. Dark green skin and 9-inch-long cucumbers. I like to harvest them around 7 to 8 inches.

Satsuki Midori. This Asian cucumber will tolerate part shade and produces long slender fruit up to 16 inches long. Nonbitter with thin skin, it's perfect for slicing.

Straight-Eight. Grows to about 8 inches long, as the name implies. This All-American Selection winner has been popular since its introduction in 1935. Great for slicing and resistant to the mosaic virus.

Tasty Jade. Another great Asian cucumber variety, Tasty Jade grows 10 to 11 inches long on huge vines grown best on a trellis. Harvest the tender fruit only 55 days after planting.

White Wonder. An early maturing cucumber that will be harvestable in about 60 days. An albino-white cucumber, the White Wonder grows to about 6 inches long and has a wide, oblong shape. White Wonder has also been called Ivory King, Jack Frost, and Albino.

Garlic (*Allium sativum*)

Garlic is one of those plants I think a lot of gardeners overlook, but it's a must in our garden. Delicious seasoning in the kitchen and incredibly beneficial for health, garlic is something we use on a near-daily basis at our house. Growing garlic is easy and when you save your own seed stock you'll find yourself with a strain uniquely suited for your local area.

Garlic grows best when it's planted like the bulb that it is—in the fall at the same time you'd plant your daffodils. While you can plant in early spring, the bulbs will be much smaller. Plant in the fall when you start taking everything else out, and let it grow until you add the last of your tomatoes and eggplants at the start of summer.

To sow your garlic you won't start with seeds, but rather with seed stock of whole garlic heads. Break the heads apart into individual cloves and plant them 2 to 4 inches deep with the root side toward the ground. Loosen the soil in the row around the garlic, making sure it's easy to work, so the heads will grow as large as possible.

Mulch over in the winter to protect from the heaviest freezes; however, most garlic varieties will survive through zone 5. In the spring the garlic continues growing until summer when the leaves begin yellowing. The yellowing leaves are your cue to harvest.

Dig up the garlic cloves and let them *cure* for two weeks. There are two main types of garlic: softneck and hardneck. Softneck garlic can be braided because it doesn't put out a central flowering stalk. Hardneck garlic often has a stronger flavor and will produce an edible flowering stalk. Hardneck garlic doesn't tend to store as well, but softneck garlic can store up until the following spring.

DEFINITION

Curing a vegetable means letting the skin dry after picking so it will last longer in storage without rotting. Vegetables that generally benefit from curing include potatoes, onions, garlic, and winter squashes.

The flowering stalk, or seed scape, of hardneck garlic will grow in a large spiral around the leaves, and then end with the flower, which forms bulblets. These can be planted and will produce additional garlic, which is a nice way to increase seed stock supply in a pinch, or cut off the flowering stalk for slightly larger heads. For most gardeners you'll want to cut the scapes off and use them in the kitchen so they don't go to waste.

Garlic shares diseases that plague onions, too, so watch out for things like onion white rot or rust. Garlic cloves may also rot if spring is very damp and rainy. A virus called yellowing virus can also spread in stock that isn't clean, so it is often best to start with certified clean seed stock. Rotating crops in the garden area can also help with all these problems.

German Extra-Hardy. Its high sugar content makes this an excellent garlic for roasting. A large and cold-hardy hardneck.

Inchelium Red. A mild-flavored softneck garlic that is considered one of the best-tasting garlics.

Polish. A hardy, softneck garlic with large, flavorful cloves.

Siberian. A cold-hardy, hardneck garlic with purple cloves. A popular hardneck variety.

Onions (*Allium cepa*)

Onions have a bit of a reputation for being picky and high-maintenance in the garden. When you prepare your beds nicely and plant the onions in loose, organic-rich soil, you'll have a better chance of success. Be sure to match the variety you plant with your garden region as well.

Large onion bulbs are easily grown in the home garden, cured until the skins are dried, and then stored for use over several weeks.
(Photo courtesy of Baker Creek Seeds)

Plant onion sets or starts about 4 inches apart with the roots just barely in the soil first thing in the spring. If starting from seed, begin onions indoors in midwinter so they will be ready to set out in early spring. Onions are rather short-rooted, so they won't compete well with weeds or tolerate heavy periods of drought well.

Onion tops can be harvested straight from the garden whenever you need green onions for seasoning or flavor. When the onion plant creates the green onion tops, they will put their efforts into the bulb creation. As the onion matures, the tops of the onions will turn brown and fall over.

Cure the onions by laying them in a sunny, protected place until the outer skins toughen. *Long-day onions* tend to form larger bulbs and store longer over the winter.

DEFINITION

Long-day onions form bulbs when the days are longer than 15 hours, so they are a good choice for southern gardens. In contrast, *short-day onions* produce bulbs when the days are 12 hours or less, so they are better suited for northern gardeners. Intermediate varieties fall in the middle.

Onions are prone to mildew and rotting if the ground is too wet during the time when the bulbs are growing. Overcrowding can also invite diseases in the onions. Onion flies lay eggs on young onion leaves or in the soil around the plants. Then the larvae eat through the stems, bulbs, and roots, so rotate crops to help break the cycle.

Ailsa Craig. Huge long-day onion that produces giant, softball-sized bulbs. The yellow flesh is mild and flavorful.

Red Burgundy. Glossy red onion with pink-tinted flesh. This onion stores well and makes a good slicer.

White Portugal. Introduced in the late 1700s, this onion has medium-shaped onions and sweet, white flesh. It's maintained its popularity for so long for good reason.

Peppers (*Capsicum annum*)

Peppers are one of those plants that everyone thinks of when they think of growing a garden. Sweet peppers and hot peppers grow in much the same way and both, like tomatoes, enjoy the warmth of summer. Peppers have many uses in the kitchen, so it's worth slipping these into the garden.

With hundreds of varieties available, you're sure to find a handful of pepper plants that do well in your particular garden space. Plant several varieties your first couple years and try saving seeds (see Chapter 10) from the plants that perform the best to develop strains suited to your region.

Start your seeds indoors so that pepper plants have a head start when the weather warms up. Peppers can take up to 25 weeks to mature, so starting plants early will help gardeners in short-season areas enjoy a good harvest of peppers before cold fall weather sets in. Transplant pepper starts after hardening off into fertile soil with plenty of compost mixed in and mulch to keep the soil consistently moist.

HARDENING SEEDLINGS

When transplanting the plants you've started from seed indoors, a hardening off period will help the plants transition. With warm-season plants, if you move them from a sheltered, indoor location straight outdoors, your plants may die from the shock. Expose the seedlings to the outdoor environment in a sunny location protected with excess wind, for a couple hours per day. Slowly work your way up to a few hours each day by increasing the time of exposure until you're ready to plant into the garden.

Space your pepper plants 12 to 18 inches apart. Provide supports, stakes, or cages as needed to prevent branches from breaking and to keep the fruit off the ground. Water regularly throughout the growing season and add fertilizer mix as a side-dressing. Peppers can be grown in containers, too, as long as they are not allowed to dry out.

Pick your sweet peppers when they are not quite fully ripened to encourage more fruit if you still expect warm weather. Otherwise let the peppers ripen for the fullest flavor or for saving seeds. Chiles and hot peppers are best ripened on the vine and care should be taken when harvesting the hottest varieties as they can burn the skin.

Common pests are usually the same as those for tomatoes—aphids, spider mites, and tomato hornworms all feed on the leaves. Blossom end rot can also occur when the plants are stressed by lack of watering or irregular water and in plants that are calcium deficient.

Anaheim. A classic hot pepper that matures in about 80 days. Fruits are 7 inches long and turn to bright red when they are mature. Bears fruit throughout the entire season.

Cherry sweet. Gorgeous fruits are cherry red and 1-inch-long, roundish peppers. This variety is more than 150 years old. Use this sweet pepper in pickle recipes and sweet salsas.

Golden marconi. These 7-inch peppers are a bright, golden yellow heirloom variety from Italy. Sweet pepper.

Habenero. A classic hot pepper for salsa and other Mexican dishes, the hot pepper matures in 85 days. Starts green and matures with orange-red coloring.

Purple beauty bell. A true purple pepper that matures in 70 days. Beautiful color on a prolific vine.

Sweet banana. A popular heirloom with long, yellow fruit. We get dozens on a single plant that mature in about 70 days.

Potatoes (*Solanum tuberosum*)

One of the most productive vegetables for a small farm space. Whereas cereal grains like wheat take a lot of space to produce usable amounts, potatoes can produce 300 bushels or more on a single acre. Even 75 feet of potatoes (planted in three, 25-foot-long rows at our house) will produce more than 500 pounds of potatoes! You can expect that 1 pound of seed potatoes will produce anywhere from 10 to 20 pounds of potatoes or more.

Potatoes create an abundant harvest in a small amount of space. This 5-gallon bucket is overflowing with bounty from a single row.

Starting with certified, virus-free seed stock will help improve yields. When you save your own potatoes to replant in following years you will decrease the amount produced. Potato seeds aren't seeds at all, but rather small potatoes, or chunks of potatoes, that are planted in the ground.

Potatoes for planting should not be from the grocery store as they will have been treated with anti-sprouting chemicals. Small potatoes with three eyes or less should be planted whole. Larger potatoes over 2 ounces and more than four eyes can be cut into pieces with each piece having at least two eyes and an ounce of flesh.

Each chunk of seed potato has at least three eyes, so it will grow well.

Dig a trench and plant your potato chunks 9 to 12 inches apart at about 4 inches deep. You want your potatoes to be planted about two weeks before the last frost so that the frost-tender vines will emerge to perfect spring weather. As the vines grow, be prepared to hill the potatoes in, or dig the dirt up around the plants.

The potatoes form on root nodules above the seed potato you planted, so as the vines reach at least 4 inches tall, dig them in. Each week, walk down the rows with the hoe and cover an inch or so of the vine with soil. Never cover more than a quarter of the vine at any one time, but do generously cover the vines and build up your potato hills.

Covering the growing vines with soil will give the potatoes more room to grow. It also makes sure that the sun is blocked from reaching exposed potatoes, which will make them green and inedible. Continue hilling up the potatoes as the vines grow until the vines stop growing and begin to fall over. We make sure to put a layer of loose mulch at this point just to help block weeds and sun.

When the vines turn brown and die the potatoes are ready to harvest. Dig your potatoes carefully to keep from cutting them accidentally. Any potatoes that you cut with the shovel are the potatoes you'll want to put in your kitchen bin for eating fresh.

Potatoes for storing need to be cured before going into long-term storage over the winter months. Spread the potatoes out in one layer on a sheet or newspaper (we use feedsacks ripped open and laid flat) until the skin is dry.

My oldest son digging potatoes last year. The joy of harvesting potatoes becomes a family affair.

Potatoes share pests with tomatoes and peppers as well as having pests unique to them. Watch out for slugs and wireworms, which attack the tubers themselves. Mineral deficiencies affect the foliage and thus the total plant production. Other diseases that affect potatoes are viruses, nematodes, and scab. Excess lime will cause scab, but starting with certified seed stock will help minimize other diseases.

All Blue. A midseason potato with blue skin that holds color when cooked. Medium-sized potatoes mature in about 80 days.

Kennebec. A late-season, white-fleshed potato that is highly useful in the kitchen. Kennebec is a great winter keeper that resists viruses.

Russian Banana. An unusual Russian heirloom potato matures to a small, tender fingerling potato. The yellow skin and yellow flesh potato is highly flavorful and stores well during the winter.

Summer Squash (*Cucurbita sp.*)

Yellow crooked neck squash and zucchini are just two types of summer squash we make a regular part of our summer garden. Summer squash is picked when the fruit is still immature so the skin is thin and edible. Summer squash varieties don't store well like winter squash, so they are eaten fresh when harvested or stored via canning and freezing.

Sow squash seed directly outdoors when the ground is warm—right about the last frost date in your area. You can start seeds indoors but use biodegradable pots so you can transplant without disturbing the roots of the plants. Plant the seeds individually 1 inch deep and space them 1 to 3 feet apart depending on the size of the mature vine and whether it's a climbing or bush form.

When given fertile soil and plenty of moisture, squash will grow like a weed. You can see the growth each day when you check the garden in the morning! Squash flowers are male or female flowers and only the female flowers will produce fruit.

Having a variety of squash plants can help fertilization because male flowers appear a little before the female flowers. The female flowers have a miniature fruit underneath the flower. Male flowers have no miniature fruit visible and the flowers are edible.

Most of the summer squash varieties should be harvested at the 4- to 6-inch range. This means you'll need to check your squash vines every day because a fruit that seems to be barely forming one day will suddenly be ready to harvest the next. Pattipan or other specialty squashes may have different sizes of maturity, but all summer squash are harvested before the seeds begin to mature.

Squash is prone to pests and grows through the warm season when creepy-crawlies are out and about in full force. If we notice an infestation of squash bugs we'll run the chickens through the garden like a flanking cavalry to demolish the enemy. I've heard from others that their chickens won't eat pests but their Guineas will, so you may want to experiment and see what works best for you. Don't keep your poultry in the garden for too long, though, or they will peck at and destroy the plants you're trying to grow.

Other organic methods of controlling squash beetles include hand picking (spray the plants with water first and it will be easier to catch the bugs) and floating row covers to prevent eggs from being laid. If you cover the plants with row covers, the flowers will have to be hand-pollinated.

Other pests include aphids, cucumber beetles, mites, and squash vine borers. Viruses and powdery mildew are diseases that can plague squash plants as well.

Black Beauty. A very dark green zucchini that is nearly black in color, the Black Beauty is a classic summer zucchini. Matures in 60 days.

Early Prolific Straightneck. This variety has been popular for almost a hundred years and won the All-American Selection in 1938. Lemon yellow squash with firm, flavorful flesh, the straightneck is highly prolific and tolerates heat well.

Peter Pan. A cute summer squash with flattened, patty pan–style fruit. The light green squash matures quickly, in about 49 days, and this squash was an All-American Selection winner in 1982.

Tomatoes (*Lycopersicon esculentum*)

The very definition of the ideal summer produce, the tomato is botanically a fruit even if it's usually grown in the vegetable garden. There is no way to duplicate that fresh, home-grown tomato taste and I think that's why tomatoes are the most commonly grown vegetables of home gardeners. Tomatoes are grown as annuals in most gardens, because they are so frost-tender; however, they are actually perennial plants.

Heirloom tomatoes allow you to grow a variety of colors, shapes, and sizes. These are several of the varieties available to the heirloom gardener.
(Photo courtesy of Baker Creek Seeds)

Tomatoes do well with soil that's been amended with plenty of compost and organic matter. Using high-nitrogen fertilizers can cause excess leafy growth and not as much fruit, so building a balanced soil is the best tactic. Tomatoes need about 150 frost-free days to set ripe fruit, and most gardeners will transplant started plants to get as much of a head start as possible.

Tomatoes transplant easily and can actually develop stronger root systems from being transplanted. Start seeds indoors six to eight weeks before the last frost date. Keep the seedlings protected from chill and when seedlings have two sets of leaves, transplant each into an individual pot, covering one-third of the stem with soil. New roots will develop along the stem and help strengthen the plant when you move it out to the garden.

Determinate tomatoes tend to form a more compact bush form and will set fruit earlier. Determinate tomatoes also set most of their fruit all within a short time frame, which makes it easier for things such as canning and making sauce, but can make a less-flavorful tomato. Many gardeners pinch off the groups of flowers growing right beside the main stem, called suckers, to encourage more energy production to the remaining flowers and fruit.

Indeterminate tomatoes (typically vining plants) do not flower all at once, and thus do not set their fruit all at the same time. This is why indeterminate tomatoes often produce more flavorful tomatoes, even if suckers are not pinched off the vine. When you plant indeterminate tomatoes, you'll want to be prepared for tall tomato vines—ours can reach more than 8 feet and we often trim the side branches to keep the plants from becoming overcrowded.

OVER THE GARDEN FENCE

Tomato plants should be kept up off the ground to prevent the tomatoes from getting mushy and rotten. Determinates tend to have a natural lift to their branches and often do well with a simple cage. Indeterminates can often have heavier branches and a more vining habit, and we've found that our plants need much more substantial support than the small cages.

Harvest tomatoes when they are ripe and the right color for the cultivar you planted. When frost is predicted, harvest any full-size green tomatoes you have left on the vine and store them in a dry, cool location where they will slowly ripen over the winter. Use tomatoes fresh, or freeze them to use later in stews, sauces, or salsa recipes.

Tomatoes can be plagued by tomato hornworms—large green worms that eat the leaves. These can be picked off by hand unless you see white, ricelike grains on the back of the worm. These white cocoons hold parasitic wasps that will emerge, killing the host, and flying off to lay eggs on other garden pests. Parasitic wasps are beneficial insects that help control the populations of these pests.

These white, ricelike cocoons are the pupae of parasitic wasps. Here you can see a newly emerged wasp leaving to kill more hornworms.
(Photo courtesy of Cindy Funk)

Colorado potato beetles and aphids are other pests that feed on the plant. Control both through beneficial insects, hand picking, and spraying leaves with insecticidal soap or even soapy water. Leaf molds, mosaic virus, and wilt are all worse in overcrowded conditions or when plants sprawl onto the ground instead of being supported off the ground. Blossom-end rot is usually a sign of soil that is not healthy and lacking in calcium.

Amish Paste. A great producer in the backyard, these 7-ounce, red tomatoes are great for paste and sauce.

Cherokee Purple. A purplish-red heirloom tomato that has enjoyed a renaissance in popularity. The large fruit matures at about 10 ounces and gets raves at taste-test competitions.

Radiator Charlie's Mortgage Lifter. Endearing for the story behind the tomato's development (see sidebar), this tomato produces a huge beefsteak fruit. Indeterminate and disease resistant, the tomatoes are highly prolific.

OVER THE GARDEN FENCE

Charlie Byles, the creator of the Mortgage Lifter tomato, sold so many of his seedlings for $1 each, he was able to pay off his home's mortgage just by selling tomato plants. His breeding goal was a large, low-acidic tomato with lots of flavor and few seeds. And he succeeded so well people were willing to pay what was in the 1930s a premium price!

Yellow Pear. This tomato has been grown for hundreds of years and for good reason. The disease-resistant plant produces tons of yellow pear-shaped fruits that are 1 ounce in size.

Winter Squash (*Cucurbita spp.*)

Unlike summer squash, winter squash is allowed to fully mature on the vine and will develop thick skins that store well for several weeks. The winter squashes include pumpkins, acorn, and spaghetti squash. They are called winter squash even though they are grown during warm weather, because the tough skins are so well-suited for long-term winter storage.

Squash can be sown directly into the garden when the soil temperatures warm up to about 55°F. Seedlings dislike being transplanted, so if you start seeds indoors you'll want to plant them in individual, plantable pots that will allow you to move them out to the garden without disturbing the roots.

Winter squash vines tend to grow very vigorously and large—it takes a lot of leaves to produce such heavy yields of large fruit! Provide your winter squash plenty of room to spread, even when growing them vertically, or they won't produce as much as they should.

Winter squashes are available in a wide variety of types, shapes, colors, and sizes. This Galeux d'Eyesines, for example, has salmon-colored skin with unique warts caused by the sugar in the skin. The flavor is almost as sweet as a sweet potato, but with a smoother texture.
(Photo courtesy of Baker Creek Seeds)

Squash vines should never be allowed to dry out completely. Even slight wilting is a sign of an overstressed plant. Mulching your plants is necessary in areas with hot, dry summers, such as where I live.

Plants can take between 12 and 20 weeks to produce mature fruit, so check the information for each cultivar carefully and make sure you plant varieties that will perform well in your region. Warming up the soil where you plant your seeds can be done with a cloche so you can get the most warm-weather growing time as possible.

Harvest the fruit when they are fully mature and the skin is thick and hard. There should be a hollow sound when the fruit is thumped, and if the cultivar has a specific color, you should be able to see this color fully developed. Cut the fruit from the vine but leave a couple inches of the vine on the squash as a stem. Squash that you plan to store should be rotated gently on the vine so that all sides are exposed to the sun and the skin thickens and hardens all the way around.

If the fruits are too large to trellis, place a board under the squash to avoid rotting. If you're aiming for a record-breaking large pumpkin, remove all the fruits except three or four early in the season. Otherwise let the fruits develop naturally and enjoy a bountiful fall harvest that will store nicely for as much as six months or more given good air circulation. See Chapter 19 for more information about root cellaring and storing food.

Winter squash shares many pests with summer squash and they can be combated in the same ways. Squash borer seems especially drawn to pumpkin vines and will tunnel through the stems, destroying an entire runner. If you see signs of a squash borer, an entrance hole with sawdustlike debris, slice vertically up the stem until you find the grub and destroy it. Cover the slit stem with a shovelful of dirt and it may reroot, preserving the growing fruit on the vine.

Amish Pie. An heirloom squash with tasty orange flesh that's great for pie. The large fruit matures at about 70 pounds.

Green Hubbard. Matures in 95 to 115 days. Dark green squash grows 10 to 15 pounds with a sweet, golden flesh. This squash stores well and is an excellent choice for winter storage.

Seminole. Developed by Native Americans in Florida, try growing Seminole in areas that are hot and humid. Easy to store, the squash will keep for several weeks—probably through the entire winter.

Kitchen Herb Gardens  9

No homestead, even on the smallest scale, is possible without a kitchen herb garden. From the earliest known examples of single-family gardens, herbs were grown for medicinal and culinary purposes—and often both at the same time. Even the makings for a soothing cup of tea were grown right at home, and there's no reason not to be able to do the same on your backyard farm.

Medicinal Herbs

Herbs that are used for health and wellness have a time-honored tradition on the homestead where self-sufficiency is a default setting. In this section I cover herbs that are grown primarily for medicinal uses. While many culinary herbs such as thyme and rosemary also have healthful benefits, this section focuses on herbs that aren't typically used in the kitchen.

See Chapter 20 for information on creating herbal preparations from home-grown plant materials.

THORNY MATTERS

While herbs can impart benefits for most people, there is a time to seek professional medical advice. This section is the barest introduction to herbal medicine and is in no way intended as medical advice, only a discussion of historical uses and current research.

Calendula (*Calendula officinalis*)

This beautiful and cheerful plant has the common name pot marigold because of the marigoldlike flowers. Easy to start from seeds, you can direct sow at the last frost date or start indoors before transplanting. Calendula thrives with plenty of sunshine but appreciates a touch of shade in the heat of the afternoon. It will not tolerate high heat, so harvest flowers early in the summer if your region is prone to extreme summers.

Calendula is a *hardy annual* that self-seeds with ease. This habit to produce volunteer seedlings makes it well suited to a prairie grass plot, cottage garden, or other informal growing area. Almost any soil will do as long as it drains well.

DEFINITION

Hardy annuals are plants that will often return the following year from self-sown seeds. It isn't the parent plant that survives, as with perennials, but rather new plants that appear from seeds that were produced from the first plant and survived over the winter.

If overcrowded, calendula can be susceptible to powdery mildew. Slugs and aphids can pester tender young plants but are rarely a serious bother.

While the flowers are edible and can be added to summer salads, the herb is most often used for wellness as a poultice. It is effective for treating skin irritations of all kinds, and that is one of calendula's primary uses.

Beginning in late spring when the flowers appear, you can begin harvesting the blooms. Calendula flower heads can be used fresh, or preserved through drying, freezing, or infusing in oil.

Chamomile (*Chamaemelum nobile*)

Chamomile is one of the most common herbal teas and is an easy herb to grow in the home garden. Informal growth habits make it a perfect plant for cottage garden border or kitchen herb container. Both the leaves and the flowers are used in herbal preparations and potpourri.

Chamomile is a hardy annual so it reseeds itself and usually comes back each year. The flowers are small and daisylike in form while the leaves are fine-cut and fernlike. Chamomile can be grown easily from seed and will also spread from rooted runners. In fact, in some areas it is considered a weed but homesteaders know better because we have a good use for the plant.

Harvest chamomile flowers by cutting the stems and drying them. Dry the flowers in a well-ventilated area out of the sun or hang the stems upside down to dry. The flowers are used to make a soothing tea that is designed to help calm nervousness and soothe stomach unease.

Comfrey (*Symphytum officinalis*)

Comfrey is an herbaceous perennial with coarse leaves that is hardy in zones 4 through 9. It grows 4 to 5 feet tall and 2 feet wide. The large root system of comfrey is extensive and makes the plant not only tolerant of drought situations, but also hard to eliminate from a garden area after it is well-established.

Comfrey is usually propagated not through seeds, but through root cuttings and divisions. They are easy to propagate and live a long time in many conditions. Many gardeners will grow comfrey in raised beds or containers, and it's often best suited to the back or middle of the perennial border.

Historically, comfrey was used both externally and internally; however, new research indicates liver toxicity when ingested. It has anti-inflammatory properties that are used in poultices for sprains, strains, and bruises of all kinds. Comfrey infused oils are also used in beauty treatments like eye creams and burn creams.

Echinacea (*Echinacea purpurea*)

Echinacea is a beautiful daisylike flower that is similar to black-eyed Susan and can be substituted for them in the landscape. Echinacea is hardy in zones 3 through 8. After it is established, you'll have flowers in two to three years.

Growing echinacea is relatively easy, as it is drought tolerant, low maintenance, and adaptable. Echinacea grows best in full sun but will grow even in part shade. The only kind of soil that may cause problems for echinacea plants would be heavy clay or overly fertile soil that holds too much moisture.

If you are growing echinacea from seeds, it will usually germinate better with a chilling period of two weeks. You can direct sow seeds shallowly in moist soil and allow them to grow right where you want them. Echinacea will naturalize readily and can be divided every three years. Use the roots from the divisions for medicinal purposes and always replant enough of the plants to keep your stand healthy and growing.

> **OVER THE GARDEN FENCE**
>
> The best time to harvest the roots is when the plant is dormant in the fall. Dig deeply to harvest as much of the deep roots as possible, because echinacea plants grow very deep roots. Most gardeners try to leave some of the smaller roots from the outside edges of the plant to allow the plants to reestablish themselves.

Echinacea extract has been shown to have immune-boosting properties in several studies, but appears to be more effective in higher doses. While echinacea grows wild through much of the United States, commercial harvesting is a threat to wild populations. Growing your own is a way to preserve the herb, and is another way to increase your self-sufficiency.

Echinacea purpurea is the most readily cultivated and is one of the three species used in herbal preparations. It is most often used in a tincture or decoction form and it is the root or leafy tops of the echinacea that is used for medicinal purposes.

Lavender (*Lavendula*)

One of my all-time favorite herbs, lavender was the only one of my newly planted perennial herbs to survive the drought in summer 2011. Lavender's tolerance for drought and poor soil is legendary for a reason, and this makes it an excellent choice for container plantings.

The main thing to remember with lavender is that it cannot tolerate wet, soggy roots, so make sure to provide good drainage. Lavender grows best in zones 5 through 8 and has a classic fragrance with purple flowers and silvery green, needlelike leaves. See Chapter 25 for an herbal craft using lavender flowers.

One of my favorite herbs, lavender is most often purchased as a started plant.
(Photo courtesy of Brannan Sirratt)

In edible landscaping you can use lavender as an attractive herb hedge or container urn. The herb combines nicely with other plants and flowers making it a rock star in the mixed border. Prune back woody growth every year or every other year to keep the plant bushy and flowering as prolifically as possible.

You can harvest the lavender blooms when they are supple, brightly colored, and fresh. It's best to harvest during a cool, dry time of the day because heat will cause the plant to release more of the essential oils. You can dry the herbs by hanging them in a dry place, or spreading the lavender flowers on a screen and placing it out of direct sunlight.

While lavender is used in both calming teas, and in gourmet dishes, it isn't a common kitchen spice. The main use of lavender is in potpourri and herbal crafts, as well as medicinal uses. Lavender is very cooling and the essential oil has been used to treat heat stroke, migraines, and muscle spasms of all kinds. Lavender also has a calming effect and is good for helping to relieve sleeplessness.

Lemon Balm (*Melissa officinalis*)

Lemon balm is an herbaceous perennial that has a deliciously fragrant citrus scent. The herb is considered invasive in some areas but will naturalize to form an attractive ground cover and is hardy to zone 4. The plant will grow up to 2 feet tall but can be pruned and pinched back to encourage bushy growth.

Lemon balm is an easy-to-grow herb that is soothing and adds a citrus taste to foods and teas.
(Photo courtesy of Brannan Sirratt)

Gardeners can sow seeds directly, or put in transplants that will naturalize throughout the garden space. If you are concerned about the potential to spread, plant lemon balm in pots or containers. It's also one of the few plants that seems to tolerate dry shade and will establish itself in an orchard or perennial garden as a hardy ground cover. The attractive flowers and sweet fragrance are a boon in the herb garden.

Harvesting lemon balm is an easy affair because it's hard to overprune or overharvest this forgiving plant. Cut back the plant two or three times during the growing season if you want to harvest large amounts all at once. Alternatively, you can harvest as needed and cut only about one quarter of the stems at a time.

As a culinary herb, lemon balm is most often used to flavor teas and alcoholic beverages. It is also used to enhance the lemony flavor of dishes such as fruit, salad, or meat. The herb is preserved by drying the herb.

For wellness uses, the herb is mostly consumed in teas that are designed to be soothing. According to *Herb Companion Magazine,* lemon balm has been shown to have antiviral, antibacterial, and antispasmodic properties. The calming effect is soothing to anxiety emotionally, and to stomach cramps and tension physically.

Stinging Nettle (*Urtica dioica*)

Nettles are often considered weeds when found growing wild in a garden, but homesteaders know better. Stinging nettles are hardy in zones 5 through 9 and will tolerate both sun and shade with enough moisture. Nettles will naturalize through seeding and cuttings and in some areas are considered slightly invasive.

To seed in the garden, or container, scatter the seeds into the soil and press them gently to make contact with the soil. The seeds are light sensitive, so don't bury them under deep layers of soil. Nettles can be dried to preserve after harvest.

Nettles are highly nutritious and historically used as a spring tonic. It is also used as a *potherb* much like spinach but needs to be boiled or blanched to take out the sting. The plant has high levels of vitamins A and C and is also high in calcium and potassium.

> **DEFINITION**
>
> A **potherb** is a leafy herb or plant used in cooking as greens or seasoning.

Stinging nettles have an antihistamine action, making them useful in relieving symptoms of allergies, arthritis, and other conditions. They are most frequently used to promote the health of the reproductive system and prostrate health. Secondary uses include strengthening the respiratory, digestive, and circulatory systems. Their diuretic action means that they should not be used by people with certain medical conditions.

> **THORNY MATTERS**
>
> Nettles are called stinging nettles because of the hollow hairs that cover the plant. When the plant comes in contact with bare skin, the hairs create skin irritation. Use gloves when handling the nettles until after they are blanched or you may find yourself paying the price!

Culinary Herbs

The following herbs are most often used as spices or seasonings to add zest and flavor to dishes. Many also have health benefits.

Basil (*Ocimum basilicum*)

Basil is a tender perennial plant that is most often grown as a tender annual. It will die at the first sign of frost and in cold climates is usually grown as an indoor or greenhouse plant. The key ingredient in pesto, basil is also used in a variety of other dishes in the kitchen. It makes a beautiful container plant option as well, with the variety of colors available.

Basil and other herbs planted in the top of the strawberry jar.

Seeds can be sown under a row cover or cold frame in late spring, or start seeds indoors anytime. You should see your seedlings appear in 5 to 10 days if you are starting the seeds in warm soil. Take special care in hardening off the seedlings you decide to move outdoors, as basil can be sensitive to cold spells.

Basil prefers a sunny spot that is sheltered and plenty of water. In order to preserve the purest flavor of unique cultivars, you should pinch off the flower spikes as they appear. This will encourage bushy growth and keep the basil directing its energy toward the leaves.

Of course if you're like me, the summer gets away from you and you suddenly realize your basil has practically gone to seed. You can cut it back, use the prunings in a salad or soup, and call it good with no harm done.

Basil grows 18 to 24 inches tall and the glossy leaves are oval shaped with a variety of colors depending on the cultivar. Basil will need plenty of moisture during periods of high heat. If you grow the herb in a container, don't let the soil dry out. Start a new set of seeds every few weeks through the summer so you can harvest successively and enjoy a bounty of basil throughout the year.

Use basil in your favorite pesto recipes—pasta salads or other Italian dishes, and stir-fries or other Asian-inspired dishes. Freeze or dry your extra basil so you'll have some when it doesn't grow well.

Caesar Basil. Strong flavor in the highly aromatic, large leaves.

Dark Opal Purple. A 1962 All-American Selection winner used to flavor vinegars and brighten pestos and salads.

Dwarf Greek Basil. This basil is a dwarf type that forms a round, bushy plant less than 1 foot tall and wide. The fragrant and flavorful leaves are also smaller than normal leaves.

Holy Basil. Purple and green coloration comes together in a highly attractive plant. The holy basil has a full and mint-tinged flavor.

Lemon Basil. A delicious, citrus-tinged flavor is produced by this basil variety and makes it popular in many dishes.

Licorice Basil. This large basil matures more than 2 feet tall. The purple, elongated leaves have a licorice flavor making it great for a variety of meat and salad dishes.

Siam Queen. A 1997 All-American Selection winner with a strong licorice flavor and large, aromatic leaves.

Cilantro (*Coriandrum sativum*)

Cilantro is the life of the party when it comes to kitchen herbs. The zest behind many Asian and Mexican recipes, cilantro is the herb form of *Coriandrum sativum*. The seeds are used as the spice coriander.

Cilantro has zesty leaves that are a must for Mexican dishes.

Cilantro seeds can be scattered in the planting area in the fall, or you can start them indoors by planting them ¼ to ½ inch deep. Give seedlings about 12 inches to spread. If you are transplanting cilantro, do so earlier rather than later, as they don't like to be disturbed after they are better established.

Cilantro reaches about 2 feet in height and will bolt to flowering when the temperatures reach 75°F for several days. Clipping the leaves regularly can help slow bolting, but after it begins setting to seed there isn't much you can do to stop it. In hot climates, offering the cilantro some afternoon shade can help prolong the ability to harvest the fresh leaves.

When the plants go to flower, you can let them—they are highly attractive and the seeds are valuable in their own right. As the flowers set seed you can cut the stems, hang them upside down with the flower heads wrapped in paper bags, and shake the seeds loose as the plant dries. The pungent seeds add a zing to salads, meat, and other dishes, but you'll want to store them in an airtight jar so bugs won't get in.

If some of the seeds fall to the ground, you can leave them undisturbed for volunteer seedlings. Cilantro will self-sow in many climates and provide plenty of seasoning in future years. Planting your cilantro crop from started seedlings purchased at the nursery won't save you very much money, if any. But planting successive sowings of seeds and allowing the summer flowers to set seeds each year in the herb garden saves you a small fortune.

OVER THE GARDEN FENCE

Fresh leaves never last long in the garden, so you have to preserve the harvest when you can. Try freezing cilantro in zipper-lock storage bags. Like basil, cilantro freezes better than most herbs.

Dill (*Anethum graveolens*)

A member of the parsley family, dill is usually grown as an annual plant. It's most often used as the main flavoring for pickles, and while it's pricy to purchase from the store, it's easy and inexpensive to grow. The fine-cut foliage and bright yellow flowers are attractive in the garden.

Dill grows tall and can be planted toward the back of an herb border. The plants attract many beneficial pollinators and insect predators to the garden. I like to include a dill plant at the end of rows in my vegetable garden, sometimes to encourage better pollination.

Sow seeds directly in the fall for spring germination, or in the spring at the last frost date. Don't cover seeds but just lightly tap them into the soil. Thin seedlings so they are growing 10 to 12 inches apart.

Harvest dill by sheering the stems and flower umbels and be sure to cut them before the flowers go to seed. After the flowers set seed, let the seed drop to the ground to provide more dill the following year. You can also harvest seeds as a digestive aid.

Dill is used not only in pickling, but also in soups, potato dishes, stews, and other dishes. Many fish recipes call for dill. Dill can be used fresh, dried, or frozen so it's easy to preserve extra harvest for use during the winter months.

Mint (*Mentha*)

One of the most recognized of all fragrances is the humble mint plant. A hardy perennial herb, mint will survive winters through zone 5 but will need to be brought indoors in colder zones. Mints grow quickly and form a low-growing mat of aromatic foliage not usually more than 2 feet tall at the most.

Mint produces flowers on little spikes in the summer, and depending on the cultivar the blooms could be pink, white, purple, or mauve. If you live in an area where mint grows naturally, you might want to plant your mint in containers to prevent it from spreading throughout the garden. Provide your mint fertile soil in well-drained but well-watered conditions.

If planting mint from seed, you'll want to broadcast seed directly in the garden area at the time of the last frost. Press them into the soil so they are no more than ¼ inch deep and water gently. Start seeds indoors four to six weeks before the last frost. Seeds take a little over a week to germinate, so don't give up on them if they don't appear right away.

Not every mint can grow true from seed, so take cuttings of those cultivars and root them. Mints are easily divided so you can take clones of the species you want. To reinvigorate older plants, you can dig them up, divide, and replant the healthiest sections.

Mint is used in so many herbal preparations—both medicinal and culinary. A given in mint teas and juleps, mint is also used in sweets, jellies, and other herbal preparations.

Apple Mint. A fruity fragrance and flavor is the undertone in this mint plant and makes it a nice choice for jams, juleps, and other sweet minty treats.

Chocolate Mint. Not a strong chocolate flavor, but the chocolaty hint is more than enough for most people to fall in love. Use this mint in any recipe where a hint of chocolate would be welcome.

Lavender Mint. This variety has red stems and contrasting green leaves. The lavender overtones are best enjoyed when this mint is dried.

Lemon Bergamot. This mint variety has a milder, lemon-infused fragrance and flavor. It's a lovely addition to tea.

Spearmint. Spearmint has a milder flavor than peppermint and is the most common mint used in sauces and fresh preparations.

Variegated Peppermint. Grown like regular peppermint, the variegated form has large white splotches on the leaves.

Oregano (*Origanum vulgare*)

Oregano is a commonly used herb in Italian dishes and sauces of many kinds. It is a strong herb, one of the more robust fragrances, and has attractive foliage.

Because it's a native of the Mediterranean region, it's no surprise that oregano tolerates poorer soil and doesn't like to keep its feet wet. Oregano is hardy to zone 5 and prefers full sun. South of zone 7 some afternoon shade will be appreciated.

Start oregano from seeds either indoors or outside. If sowing seeds directly outdoors, wait until the soil temperate is at least 55°F. Plant seedlings indoors six weeks before the last frost date and allow up to two weeks for seed germination. Transplant when seedlings are around 6 inches tall and pinch off the tallest-growing tips to encourage bushier growth. Space plants at least 12 inches apart.

Harvest oregano at any time, but expect the best flavor right before the flowers appear. Cut off the outer stems first as they will mature ahead of the growing center. Don't harvest more than half the plant at one time or your oregano will not have enough energy to survive the winter.

Use oregano fresh throughout the growing season. Store extra by freezing or drying the leaves. Sprinkle oregano leaves on salad to add extra flavor, or chop it up and sprinkle over meat on the grill.

Parsley (*Petroselinum crispum*)

Parsley is a biennial plant with attractive leaves high in vitamin C. It's a pretty plant in the garden and a nutritious herb in the kitchen. Enhance the flavor of sauces, herb butters, rice, and stir-fries with the flavorful leaves. The roots of the Hamburg parsley are also edible and similar to parsnips.

Parsley comes in flat leaf, curly leaf, and root-producing varieties. Flat leaf, also called Italian parsley, is generally more nutritious, while curly leaf parsley is used mostly as a garnish to enhance the appearance of food.

Parsley is slow to start from seeds. You can soak the seeds in water overnight to help speed up germination. Try to keep the soil at about 70°F and that should help as well. Start seeds six weeks before the last frost date so your parsley will have a good start when the weather is warm enough to transplant them outdoors.

Parsley should be harvested before flowering begins unless you want the plant to set seed. You can let the parsley overwinter under a layer of mulch and in the spring the plant will flower and set seed so you'll be able to gather the seed by late spring and set out new plants for the year.

Keep the parsley in moist soil and provide part shade when the summers are too hot. Parsley appreciates a thick layer of mulch or compost—as most of the vegetables and herbs do.

Parsley is a must-have in my herb garden. Note the beginnings of a flower cluster forming in the plant.

Parsley doesn't have too many disease problems but has some pest problems. Parsley is related to carrots so it makes sense that it is sometimes preyed on by carrot flies. Aphids or mites can also be a problem on occasion.

Harvest leaves for storing at their most flavorful peak of summer. Freezing parsley is generally a better option to preserve flavor than drying. Chop the parsley and add to water, freezing the mix in ice cube trays. If you want to dry parsley it should be done as quickly as possible without overheating.

OVER THE GARDEN FENCE

Parsley, like carrots, helps attract predatory wasps and other beneficial insects to the garden. Because of this effect, it is an herb commonly used as a companion plant in the vegetable garden. I know a master gardener in our area who always puts a parsley plant at the end of each tomato row.

Rosemary (*Rosmarinus officinalis*)

One of my absolute favorite plants, rosemary has a lot to offer a backyard farmer and a gardener. The evergreen perennial can be included in a fragrance garden, mixed perennial border, or large ornamental container. Hardy in zones 4 through 10, the bush is only evergreen through about zone 8.

Most rosemary plants have an upright growth pattern and can reach 4 or 5 feet in good growing conditions. Some cultivars, however, have dwarf forms and don't grow more than a couple feet. One of my favorite cultivars has a weeping form that is perfect for edging a raised bed or large container.

I love the pine needle–like foliage of the rosemary plants and usually have more than one variety.
(Photo courtesy of Brannan Sirratt)

Rosemary can be grown from seed but takes a couple weeks to germinate. Provide plenty of sunshine and moderate water until the plant is well established. After rosemary is well established it will prove to be more drought tolerant than many other garden plants.

Not many pests will attack rosemary, but whiteflies, spider mites, and powdery mildew can be occasional problems. When growing tips are damaged by the cold, or get too woody and overgrown, simply prune out the dead or twiggy branches. Use them in a grill to flavor baked potatoes, roasted kabobs, or grilled meat.

To harvest rosemary, simply snip fresh stems from the plant with hand pruners. As with most herbs, harvesting leaves prior to flowering is the time of peak flavor. Preserve the harvest by drying, freezing fresh twigs, or flavoring oils and vinegars. Rosemary is often used in potato dishes, meats of all kinds, and even herb butter.

Arp Rosemary. Grows to more than 5 feet tall. This rosemary has pale blue flowers and is very cold-hardy to zone 6.

Golden Rosemary. Also called Golden Rain. This rosemary has bright yellow-green leaves instead of the dark green of most rosemary plants. The flowers are pale blue.

Pink Rosemary. This rosemary variety has a slightly weeping variety and produces pale pink flowers. Prune yearly to help the shrub hold its shape.

Santa Rosa Trailing Rosemary. One of my personal favorite varieties, the trailing or creeping rosemary has a weeping habit. Plant it in a raised bed or container where the trailing habit can be enjoyed!

Spice Island Rosemary. This 4-foot-tall variety grows with a strong upright growth habit. Dark blue flowers appear in late winter through early spring.

White Rosemary. Beautiful bushy habit, this rosemary grows 4 feet tall and wide and produces pure white flowers.

Sage (*Salvia officinalis*)

Sage is a perennial plant, almost shrublike, with highly fragrant foliage and brightly colored flowers. The leaves are usually textured and often have a silvery or grey tint to them even in the colored varieties. Sage is one of those plants that many add to their ornamental gardens without realizing what a boon it is for the kitchen as well. The perennial shrubs are hardy in zones 4 through 10.

Sage can be planted in the autumn with other shrubs, or in the spring. To start seeds, sow indoors six weeks before the last frost and transplant after the weather has warmed up a bit but allow plenty of time for the seeds to germinate. Provide sage with well-drained soil because the shrub is drought tolerant and doesn't require excess water.

Give sage room to spread its wings—the greedy shrub will easily take 2 feet in the garden space. Provide sage with full sun and well-drained soil and it will provide plenty of fresh leaves for harvesting. Sage is best used fresh and you can add a few leaves over the hot coals of the grill to smoke meat.

There are few pests that attack sage, but spider mites can be a problem sometimes. Slugs can also attack sage, so watch for signs of those slimy trails. Poor air circulation can make powdery mildew more likely, so give your salvias enough room.

Because sage is a perennial, you do not have to harvest all the leaves in one fell swoop. Harvest about a quarter of the leaves each year and allow the plant to grow back each year. If you want to be fussy with your sage you can pinch back the flowering stalks, but I enjoy the hummingbirds that are attracted to the flowers.

Sage and rosemary planted together in a large ceramic container. Geranium is also planted but it won't survive the winter. Next year when the perennial plants are larger, they will have more room in the container.

Sage is great in many chicken dishes, stuffing, or vinegars. Sage should be dried thoroughly before storing but be aware that dried sage has a stronger flavor than fresh leaves. If the plant becomes too woody and the center begins dying off, you can prune it more heavily to encourage new growth. Or you can root cuttings or *layer* the plant to create new sage plants.

DEFINITION

To **layer** a plant is to pull down a growing stem and bury part of the stem under the ground. The middle of the stem is under the soil with one end still attached to the main plant, and the other growing tip above the ground. After a few weeks the stem will develop a root system of its own and can be cut off the main plant. Replant the cutting and enjoy a brand-new plant identical to the original.

Thyme (*Thymus spp.*)

Thyme is another of my personal favorites. The scent is delightful, and the plant is so easy to grow in bed, border, or container. I use thyme as a ground cover for container shrubs and have a nice amount produced each year as a result. Because the plant is semi-evergreen, the thyme is a great foil for other flowering plants and herbs.

Start thyme by plugging in plant starts or divisions in the spring or fall. You can also seed the plants directly outdoors in early spring or start seeds indoors and then transplant them after the last chance of frost. Thyme is hardy in zones 4 through 8 and is grown as an annual plant elsewhere.

Thyme is a creeping plant and will spread via runners. This tendency to form a pleasing mat makes thyme a popular choice for a lawn replacement. When allowed to bloom, thyme has attractive tiny flowers in white or pink.

Thyme needs little maintenance in the garden beyond periodic watering. After the plant matures the woody center will begin to die out. Every three or four years simply dig up your thyme, cut away the woody center, and replant the healthy green sections from the outside. Extra divisions can be replanted elsewhere or given to friends.

Harvesting thyme is as simple as cutting off a leafy stem to use fresh. In mild areas where thyme stays green year-round, you can harvest throughout the year. In other areas, save some extra harvest by hanging bunches to dry in a shaded area out of the sun.

Thyme is highly fragrant and each cultivar has a unique taste and fragrance of its own. Lemon thyme can be used in recipes that call for lemony zest, or even in teas and jelly. Standard thyme varieties are a favorite in meat dishes and seem to go especially well in Italian recipes.

Elfin Thyme. Elfin thyme only grows a couple inches tall and produces the typical pink thyme flowers, but they appear several weeks later in the middle of summer.

French Thyme. French thyme has a sweeter flavor than the common English thyme and the leaves have a silvery tint to them.

Orange Balsam. Strongly scented and flavored, this taller thyme is a workhorse in the kitchen.

Silver Thyme. A variegated thyme, the silver variety has gorgeous white edges that add sparkling color as a ground cover. Used as common thyme in both kitchen and garden, this thyme adds a new sparkle of color.

Fruits, Berries, and More  10

Fruit has a bad reputation for being difficult to grow and for needing a lot of dedicated space to be productive. But nothing could be further from the truth. Incorporating fruit, berries, and nuts into the backyard farm can be part of a healthy and thriving homestead. This chapter covers some of the easiest to grow or most common fruits, berries, nuts, and edible perennial plants that you might want to add to your backyard farm.

The best way to increase the amount of fruits you produce in your backyard is to think outside the box. Does fruit *have* to mean an orchard of apple trees? You can plant fruiting vines, fruiting bushes and shrubs, and fruiting ground covers like strawberries in myriad places that a traditional apple tree wouldn't fit. (I'm not implying that there's anything wrong with apple trees!)

Fruit on Just an Acre? Yes!

You might think that you don't have room for fruit plants, but if you have room for plants at all, you have room for homegrown fruit. Fruit on a small scale works best when you incorporate your fruit or berry plants wisely. Incorporating edibles into the ornamental landscape is one great way to do this, which we discussed in Chapter 4.

Another way to fit fruit into a smaller available area is to choose fruit plants that are more reasonably sized than a full-size 40-foot tree. Many fruit trees are available in dwarf, or upright columnar forms. These begin as short as 6 feet tall and are grafted onto dwarf rootstock to limit their growth. You could even plant these trees in large containers and provide fruit for your family as part of your patio container garden!

Some fruit trees, such as apple, pear, and plum trees, can be trained to stand against a fence or into a thinner form. This tree form is called espalier, and while it requires more frequent pruning and can limit the amount of fruit produced per tree, it is one way of increasing the number of fruiting plants on your property overall.

ON A DIFFERENT SCALE

Fruit trees are available in different sizes based on the type of rootstock they are grafted to. Here are the size breakdowns and some of the common rootstocks you might see in each size category:

- **Miniature.** Grafted onto M27 rootstock; the full size is 2 to 3 feet.
- **Dwarf.** Grafted onto M9 or M26 rootstock; the full size is 5 to 8 feet.
- **Semi-dwarf.** Grafted onto M7 or G30 rootstock; the full size is between 8 and 15 feet.
- **Standard.** These rootstocks produce full-sized trees that vary according to the fruit type.

Apples (*Malus*)

Were apples the original fruit? No one knows, but there are so many apple cultivars (varieties) available it's no wonder that people sometimes think so. With so many selections available, there are apples grown in every state in the United States and some that will tolerate from zone 3 to zone 9. Not every cultivar will perform well where you live, however, so get advice from friends and neighbors around you, or your local county extension office.

Apples are an interesting challenge in the landscape because you need a second cultivar that will cross-pollinate your trees. Sometimes you can buy specially grafted trees that have branches from more than one variety grafted into a single tree, and where room is at a premium this might be the best choice. Otherwise plant two or three different types of apples within 25 to 75 feet (closer together for the smallest dwarf types) to ensure good pollination of your apples.

THORNY MATTERS

Some apple trees produce sterile pollen. Jonagold, Mutsu, and Baldwin are three popular cultivars that will not pollinate other apple trees. If you plant one of these cultivars, you'll need at least two other trees within pollination range. Check labels or cultivar descriptions and look for any that state "infertile pollinator" for your clue!

Apples, like many fruit trees, appreciate woody mulch underneath the trees. In fact, some holistic gardeners use the growing tips of the branches that are pruned throughout the year, chip them in a wood chipper, and spread them back with compost under the fruit trees. Apple trees also need a good amount of water and plenty of sunshine to develop those large, succulent fruits!

With apples especially, backyard farmers should try heritage cultivars or those cultivars that aren't necessarily popular in the grocery store. Often the commercially popular types are those that are not disease resistant and require lots of spraying and chemical applications. Different cultivars need differing amounts of "chill hours" or time under 45°F where the cool temperatures ripen the apples and enhance the flavor. In warmer climates you should choose cultivars that are considered "low-chill" that don't need as many cool-weather days.

Apple trees with fruit that matures in the summer are usually not the best cultivars for storing through the winter. For winter storage and fresh fruit to snack on during the winter, you'll want apples that mature in late fall.

Apples can bear fruit heavily, bending branches if not thinned when fruit is small.
(Photo courtesy of Jill Browne)

Cultivars to try: *Arkansas Black* has a tart flavor considered good for cider and has a firm flesh that stores well. *Dorsett Golden* ripens early, has high yields, and only needs about 100 hours of chill time. *Hardy Cumberland* was developed by the University of Tennessee and is a sweet apple with good disease resistance. *Jonathan* is a classic heirloom from the mid-1800s that is a mid-season apple with good storage properties and full, tangy apple flavor. *Pink Lady* or *Cripps Pink* requires 200 to 400 chill hours and is a late-season apple that stores well. *Winesap* is a dark red, late-season apple that does well in the Midwest and Mid-Atlantic areas.

Sweet Cherries (*Prunus avium*) and Sour Cherries (*Prunus cerasus*)

Cherries come in basically three forms: sweet cherries, sour cherries, and less commonly grown bush cherries, which have several different varieties. Cherries are a great option for substituting with other ornamental flowering trees, such as redbuds or forsythia, to transform your landscape into an edible landscape. Sour cherries, and a few of the sweet cherry cultivars, are self-pollinating, which makes it easier for backyard farmers with limited space to add fruit trees to their property.

Cherry blooms are a gorgeous addition to the spring landscape. And of course, cherry trees go on to provide edible fruits for the backyard farmer later in the season.
(Photo courtesy of Tim Sackton)

Like most fruit trees, cherries grow best in full sun and well-drained soil. Cherries need to be kept consistently moist until they are fully established, because if the roots dry out they will die. This is especially important to note if you're trying to grow them in a container. As with apples, they do well with a thick layer of organic mulch.

OVER THE GARDEN FENCE

You won't be the only one who loves the cherries your tree is producing. Birds can clean you out! You can hang old CDs or scary "eyeballs" to try to deter them. But the best protection is to cover the bush or trees with netting as soon as the color begins to change.

You can eat cherries straight off the tree and can get as much as 25 to 150 pounds per tree. Cherry pie, juice, and preserves are all ways of storing extra cherries you've harvested. You can also freeze the cherries on a flat baking sheet and then store frozen in bags.

Sweet cherry cultivars to try: *Black Tartarian* has huge dark, almost purple fruit and tasty flavor. *Lapins* is one of the self-fertile sweet cherries and is a heavy producer with excellent crack resistance. *Royal Lee* is a low-chill cultivar that flowers early in season. *Sweetheart* is another of the self-fertile sweet cherry cultivars and is crack-resistant. *White Gold* has yellow skin and light color flesh, and is a late-blooming cherry hardy to zone 4.

The cherries on the tree are highly attractive and, of course, delicious. A true win-win in the landscape.
(Photo courtesy of Rachel Matthews)

Sour cherry cultivars to try: *English Morello* has tasty, rich-red flesh and is a low-chill cultivar on a naturally small tree. *Mesabi* is a sweeter hybrid cherry that isn't quite as tart as most sour cherries. *Surefire* is a tangy sweet cherry that blooms later than other sour cherries and is highly crack-resistant.

Other cherries to try: Nanking cherry (*Prunus tomentosa*) is a bush cherry that is about 6 to 10 feet tall with white or pale pinkish flowers. Plant two cultivars for best pollination and use them in the landscape as a hedge alternative to increase the usefulness and productivity of your landscape.

Mulberries (*Morus sp.*)

Mulberry trees may be unfamiliar to some because it's not a fruit that is available commercially. The fruit doesn't ship well and won't keep long, so enjoy it while it's ripe. The mulberry fruits look similar to raspberries and have a super-sweet taste to them. Mulberry trees are hardy in zones 5 through 8.

Mulberries are highly variable—I've seen everything from small shrubs with multistemmed forms to huge trees 50 feet tall and wide. There are even weeping forms of mulberries, which can add an interesting shape to the landscape. They are self-pollinating, so you can add just one to the landscape as desired.

THORNY MATTERS

When you plant your mulberry, don't put it in an area where it will overhang a walkway or driveway area. The dropping fruits can be messy and stain pavement or car paint.

It's not a tremendously striking tree necessarily, but mulberries are fruitful, easy to grow, and have few, if any, pests to contend with. It will sometimes bear fruit the first year after being planted. But you'll get a good crop the second year no matter when you planted the mulberry, spring or fall.

To harvest, spread a sheet or tarp (one that you don't mind getting stained) and shake the branches. Ripe berries will fall and usually the unripe berries will remain on the tree. Harvest berries every three days during the month or more as they are ripening.

You can see how the mulberries resemble raspberries. These are just beginning to turn color and will be ready for harvesting in a few weeks.
(Photo courtesy of Jennifer Harshman)

You'll want to use the berries right away because these sweet, soft fruits don't last very long. Eat fresh (my kids' favorite way to enjoy them) or make them into a wine or sauce. I've also seen them as pies or tarts and also canned as preserves. However you decide to enjoy them, the full, sweet flavor is one of the best tastes of the summer for our family.

Peaches (*Prunus persica*)

Peaches are hardy in zones 5 through 9 (some hardy to zone 4) but have a reputation for being a little more picky and difficult to grow than some of the other fruit trees. The fresh taste of a juicy peach straight off the tree can't be beat, though, so it's worth the effort in my opinion. Many of the problems with peaches come from frost.

Peaches have attractive (not to mention delicious) fruit, and beautiful stems and leaves.
(Photo courtesy of Steven Depolo)

Peaches can be planted on the north side of the home, barn, or stand of trees where winter shade will be provided so the temperature during fall and spring season changes will be more even instead of erratic. You can also paint the tree trunk and major branch joints with white paint after leaf fall to reflect the winter sun and help protect the tree from damage caused by temporary winter thaws.

There are tons of peach varieties to choose from, so try checking with your extension office for tips on which ones to try. In southern climates with mild winters, you should look for low-chill cultivars. Because peaches are generally self-pollinating, you can plant a single peach tree on your property. In fact, the gorgeous pink flowers can easily take the place of an ornamental pear that won't bear fruit for you!

Peaches come in two types: clingstone peaches in which the fruit is attached to the pit inside, and freestone peaches in which the pit is more detached and separates from the flesh better. Growing your own peaches allows you to select some of the most delicious, thin-skinned peaches from years past, the likes of which you've never purchased at a grocery store. Commercial growers select varieties with harder flesh and thicker skins that hold up well to the abuse of shipping hundreds of miles. Your peaches only have to make it from tree to teeth.

OVER THE GARDEN FENCE

Nectarines are a naturally occurring sport (wild mutation) of peaches. Nectarines are fuzzless peaches and are produced both from breeding nectarines to nectarines and from grafting sports that appear on peach trees.

Prune your peaches and nectarines to an open, vaselike habit so sunlight can reach each branch. Peaches use a lot of nitrogen, so a good dose of rich compost and mature chicken manure each year after the leaves fall will be welcome. Because most peaches and nectarines are small trees not growing larger than 15 feet naturally, they make a great tree choice for near a house or in a small landscape.

Even in the best of conditions, peaches are generally shorter lived than other types of fruit trees. Ten years is a good run for a peach tree; 20 years is fabulous. But don't let that stop you! They are small trees, not foundation shade trees, so easily replaced if needed.

Peach cultivars to try: *Avalon Pride* is a relatively new, very flavorful peach with resistance to peach leaf curl. *Contender* has a high chill requirement and is hardy to zone 4, making it awesome for northern climates as it has a later bloom time that won't get zapped by late frosts. *Harbelle* is a small tree with productive habit that bears lots of fruit and is resistant to bacterial leaf spot. *New Haven* is similar to *Red Haven* but has a much higher disease resistance and fabulous productivity. *Rio Grande* has a low-chill requirement with delicious freestone peaches and unusually beautiful flowers.

Nectarine cultivars to try: *Fantasia* is a large freestone nectarine with a gorgeous reddish skin and sweet yellow flesh that is said to be one of the tastiest nectarines. *Hardired* was bred in Canada and has excellent cold-hardiness and large blossoms in the spring. *Panamint* is a low-chill variety with a full tangy and sweet flavor.

Pears (*Pyrus sp.*)

Pears are easier to grow with organic care than apples if you choose a fireblight-resistant cultivar. Fireblight is a disease caused by bacteria that affects pears, apples, and roses. Pears are excellent as a dessert fruit and for eating fresh off the tree. I also love making pear sauce when the pears are ripe to enjoy warm over pancakes, oatmeal, or ice cream for months thereafter. Consider adding a large, accent shade tree to your landscape, or plant several of the smaller dwarf pear trees as a wall, driveway edge, or espaliered along a fence.

I've never entirely grasped the reason for planting ornamental pears, which do nothing but bloom, when you can plant a pear tree that bears actual fruit. Most pears need cross-pollination, though, so if you plant one, you should plant at least one more suitable pollination partner to ensure a good fruit crop. Pears are generally hardy in zones 5 through 9, but an increasing number of cultivars are now hardy through zone 3.

If you have an area with less than perfect drainage that holds rainwater a day or two, that might be a better place to site your pears instead of another fruit because pears can tolerate slightly wet feet. Standard pears produce more fruit but grow much larger, while dwarf pears reach less than 20 feet and begin bearing fruit sooner.

I love how prolific the blooms are on our old pear tree. This tree is at least 40 years old and still looks amazing every year.

This younger pear tree is being well trained to an attractive espalier shape. It nicely fills this otherwise odd space along the outside of the building.
(Photo courtesy of Jill Robidoux)

Pears need more pruning when they are growing rapidly than some of the other fruits, especially if you're training them to a small space or special shape. When they are established and fruiting they don't need as much pruning. Pears tend to do well trained to a central leader form and usually have a naturally upright habit, but they are easy to train into gorgeous espalier shapes if desired.

This is a fully mature pear tree that has been well maintained in the espalier. The attractive bark is gorgeous, the spring blooms hint at fruit to come, and the shape is eye-catching.
(Photo courtesy of Jill Robidoux)

As with apples and peaches, you should thin your pear tree fruit when the fruit is very small so that your trees will bear fruit each year, instead of alternating years of fruit.

Harvest the European pears while they are still a touch unripe and they will ripen after harvest. The color change of the skin is your clue to harvest. Pears can store for a while, in some cases several weeks, especially if harvested underripe and kept in a cool place.

Cultivars to try: *Blake's Pride* is hardy to zone 4, a boon for northern gardeners, and is a delicious, light-gold pear. *Comice* is a classic-shaped, large pear with a soft, sweet taste and it performs best in the West. *Seckel* is a small pear but it has a delicious, full flavor and is not only fireblight resistant, but is hardy through zone 4. *Summercrisp* is a pear known for reliable fruit production and sweet flavor that performs well in zones 3 through 5. *Warren* needs only 600 chill hours and is a large pear with smooth, juicy flesh and good disease resistance even in more humid climates, which is why it excels in zones 7 through 9.

Persimmons (*Diospyros sp.*)

Oriental persimmons (*D. kai*) are hardy in zones 7 through 10 and usually grow no larger than a medium tree. We had one in our Texas home that was only about 6 feet tall as it had been grafted on dwarfing rootstock. American persimmons (*D. virginiana*) are medium to large trees, some reaching 70 feet or more, and are hardy in zones 4 through 10.

Persimmons are rock stars in the landscape as they are gorgeous summer trees, have interesting bark in the winter, and of course, the bright orange fruits are highly ornamental in the autumn. The only season it doesn't shine is the spring, because the flowers come later in the season and are an unnoticeable green color. Most persimmons are self-fertile but a few of them do better with cross-pollination.

American persimmons should be harvested after the fruit becomes slightly soft, usually after a frost or two. Oriental persimmons can be picked before fully ripe, but after the color matures, and allowed to ripen on the counter. Persimmons can be frozen, and in many cultures they were fermented.

Plums (*Prunus sp.*)

Plums have beautiful spring flowers and delicious fruit, with easy-to-grow cultivars available in most areas, but can be variable in fruit set and disease resistance. There are European plums (hardy in zones 4 through 8), Oriental plums (hardy in zones 6 through 10), American plums, and many hybrids, and each react a little bit differently. Most plums grow small- or medium-sized naturally and often have an open, shrubby growth habit.

Spring blooms are fragrant and beautiful, and the succulent fruit is a boon to the backyard farmer. We have wild plums that grow in this area that most just call "sand plums" although I don't know what their proper scientific name is. Like apples and pears, plums do better if the fruit is thinned early in the spring. Plum branches can grow so heavy from the weight of the fruit that they break right off the tree!

Plums should be pruned to an open shape when they are young. Provide plenty of sun and clean up any old growth or sucker sprouts you find popping up in the spring. And while you should prune each year for the best fruiting results, I must admit to walking along the creek bed and picking wild plums that haven't been touched in years. That's the get-it-doneist coming out in me, though. When the plums are in my immediate yard landscape, they get a lot more hands-on care.

OVER THE GARDEN FENCE

Diseases are more common in humid areas and if plums are overcrowded. Disease-resistant varieties will help prevent problems with your plums, so check the labels and descriptions carefully. It will help to make sure you are planting plums that are resistant to the diseases most common in your particular region, so check with your local extension office to see which pests and diseases to watch out for.

Harvest plums when they come off the tree with a slight twist of the stem. Most of the plums on the tree will ripen all at once, so be prepared for pounds of delicious fruit at one time. Plums will store for a few days in the fridge, but then you'll want to preserve them somehow. You can dehydrate them, make pies or tarts, and can jams or preserves.

Vines and Berries

Fruiting vines and smaller berry bushes are one of my favorite ways to sneak deliciousness into the landscape. They are easy to tuck into off spaces—along the walls and fence lines, or into the awkward spaces between trees and perennials in mixed borders. This section discusses some of the most common berries and fruiting vines that you might want to add to your landscape.

Blackberries and Raspberries (both *Rubus spp.*)

These brambling fruits are hardy in zones 3 through 9 depending on the type and variety. Most brambles bear on second-year canes, so you should prune one third of the canes each year to keep your brambles under control. Raspberries will spread by runners under the ground, creating larger and larger thickets each year if not periodically cut back.

Trellis your raspberry or blackberry vines against a fence, arbor, or wall to make it easier to pick them. You can also choose cultivars that are thornless, which also makes it easier to pick ripe berries. In fact, the thornless varieties make a great addition to a kids' garden area because children are irresistibly drawn to these delicious summer berries.

There are few tastes that say "summer" better than the fresh-off-the-vine flavor of raspberries and blackberries.
(Photo courtesy of Rachel Matthews)

Raspberries and blackberries should be planted anywhere from 2 to 5 feet apart in the row in which you want them to grow. They will quickly grow additional canes to fill in the gaps in between, and in a couple years you'll have a beautiful and fruitful living fence.

Viruses can create problems for raspberries, but especially for black raspberries and blackberries. Start by planting certified virus-free stock but be prepared to renew your beds every 10 years or so if your brambles become less productive. Find out which varieties do best in your growing area for clues on disease resistance that will work in your favor.

Raspberries are available in the classic red colors (Autumn Britten, Killarney, and Latham), purple (Royalty), black (Black Hawk and Jewel), and even yellow (Anne) so your backyard can sport myriad jewel tones. Raspberries do not store well after harvest, so pick them and eat them promptly. After a couple of days the almost-ripe berries you picked will be fully ripe and very soft, at which time your best bet is a cobbler or canned preserves. In a pinch you can freeze the berries on a cookie sheet and then store in a freezer bag, but the thawed berries will make better sauces or cobblers than eating out of hand.

Blueberries (*Vaccinium spp.*)

There seems to be a blueberry bush that is suitable for almost everyone. Highbush blueberries (*V. corymbosum*) are hardy in zones 4 through 7, lowbush blueberries (*V. angustifolium*) are hardy in zones 3 through 7, and rabbiteye blueberries (*V. asheii*) hardy in zones 7 through 9. While most blueberries are semi-self–fruitful, they all do best with other bushes around for cross-pollination, but because they perform so well in the landscape there's no reason not to plant several!

Blueberries have pretty white flowers, beautiful fruit, and in the autumn many varieties have colorful foliage as an added bonus. The fruits are delicious and highly nutritious. Highbush berries are the ones you typically think of as blueberries, and rabbiteyes tend to be smaller even though they are very prolific. Lowbush fall somewhere in the middle and are the berries traditionally used in pies and for canning. All will provide delicious fruit for eating, baking, and storing.

Blueberry shrubs vary in size from 1 to 8 feet or more (pruning keeps them easily under control) and can be grown in rows, containers, or mixed into a woodland garden. Blueberries tolerate part shade but bear more fruit with at least six hours of sunlight each day. The main key to growing these otherwise easy-to-grow plants is providing the right soil pH. Blueberries need acidic soil in the 4.5 to 5.5 range, similar to azaleas, heather, and other woodland plants. The other main thing to watch out for with your blueberry plantings is that the soil is well-drained and doesn't hold water.

When you plant a blueberry bush, dig a generous hole, at least three times larger than the rootball you're planting. Amend the soil you put back into the planting hole with half the original soil, and half a mixture of peat moss and pine needles, oak leaves, or woody compost from those trees. If you are planting in a container, use this type of mixture as planting soil, and mulch with acidic mulches.

Blueberries grow easily on bushes as long as their soil pH needs are met. One of the most nutritious (not to mention pricey) fruits can be grown right in your backyard.

Blueberries don't need very much pruning, but as with all shrubs cleaning up dead, broken, or over-crowded branches will improve the bush and help prevent diseases. Blueberries ripen over a long period, so you'll want to pick ripe berries every couple of days or so as they turn. It's okay to skip a couple days because berries will hold well on the bush. You know they are ready to pick when they fall off in your hands with a light tug.

Fresh berries are fabulous for eating and will last for a few days in the fridge. You can also freeze them, cook with them, bake pies, or make jams. Blueberries also dehydrate well and can be added to granola or trail mixes.

Grapes (*Vitis spp.*)

Grapes are a classic arbor vine, and for good reason. They provide bold foliage for shade in the summer, decorative vines, and classic fruits. There are several varieties of grapes that will grow across the United States in zones 3 through 10. The biggest consideration for adding grapes to your backyard farm is that the vines are thick when mature and need a sturdy support, and that they need yearly pruning.

There are two types of grapes that are native to the United States: the fox grape and the muscadine grapes. European grapes do best in the West with dry, hot summers. Muscadine grapes are some-times called *Scuppernong* grapes, which is actually the name of the most popular cultivar, and the leaves of these native grapes turn yellow in the fall.

Grapes need full sun and well-drained soil to give their best fruit. Pruning the vines each year allows the grapes to stay healthy and productive. It is possible for grapes to overproduce their root systems, with too much cane, leaf, and fruit growth that ultimately destroy the plant. Grapes don't need heavy fertilizing, so just top the soil off yearly with compost and a good mulch.

This newly planted grape vine will need lots of watering the first year or two as it becomes established. After that it will need less watering, but a lot more pruning!

When harvesting grapes you usually cut the entire fruit cluster at once, and leave the grapes on their individual stems. Muscadines are the exception to this and sometimes ripen over time instead of all at once. Grapes will ripen from July through September depending on region and type.

OVER THE GARDEN FENCE

Various grapes are susceptible to black rot fungal disease, which is best prevented with good pruning and cleanup of debris. Downy mildew is a problem in wet and humid climates. And of course, as with any delicious fruit, there are pests who enjoy eating them as well. Your county extension office should be your first stop in finding out what you can expect in your area.

Pruning grapes is relatively easy. Allow one or two main branches to form, and limit cane growth from there. Each year you'll cut back the previous year's growth to about 12 to 18 inches. There are a few different ways that grapes can be pruned, and it will mostly depend on where you're growing the grape vines. Just keep in mind that you want to prune away the bulk of last year's growth in late winter or early spring and you'll be okay. Check out BackyardFarmingGuide.com for specific examples of pruning grapes.

Kiwifruit (*Actinidia spp.*)

Kiwifruit, especially the hardy vines, are pretty and productive in the landscape. They are a great option for climbing plants along an arbor, trellis, archway, fence, or wall. Kiwifruit vines have pretty foliage (sometimes variegated) and yummy, edible fruit. Kiwifruit (*A. deliciosa*) is hardy in zones 7 through 9, and hardy kiwifruit (*A. arguta*) is hardy in zones 4 through 9.

Kiwi vines need cross-pollination between male and female vines. You'll need a minimum of one male vine for every eight female vines to make sure you get good pollination. Late frosts can damage pollen and hurt fruit production for the year, so if you expect a late freeze you can protect the vines by wrapping them in burlap or piling on straw to try to save the productivity.

Kiwifruit vines will grow huge in optimum conditions so lots of pruning may be necessary. They usually grow best in full sun, but in hot climates they'll appreciate some part shade. Give your vines well-drained soil and prune yearly to avoid overcrowding, and you'll prevent most diseases. Japanese beetles tend to be the biggest pest problem, so you may set traps if you see them hovering around your vines.

Prune kiwifruit vines the second year, after they put on a good year's worth of growth. Don't be afraid to prune them back pretty good each year in late winter or early spring to just a few inches of growth from the previous year. In optimal regions your kiwi vine is capable of growing 20 feet in a single year, so you might have to prune back the growing tips during the summer as well.

Harvest kiwifruit when they are just a tad underripe and they will soften when you pick them. Kiwi will ripen quicker at room temperature but will store for a few weeks in the fridge. Hardy kiwi has thinner skin that isn't as fuzzy as the traditional kiwis you find in the grocery store, and is smaller in size.

Melons (*Cucumis melo* or *Citrullus lanatus*)

No vegetable garden is complete without a sweet, mouth-watering melon ripe off the vine. Notorious for being hard to grow and taking up a lot of space in the garden, I have success growing them vertically. In our small garden space I only grow a few vines of varieties that produce smaller fruit that can be trellised without breaking off the vine.

There are four main kinds of melons: watermelons, American cantaloupe (muskmelons), honeydew, and specialty melons such as Asian melons. Watermelons tend to resist bacterial wilt better than the others, but require between 80 and 100 days of true summer-warm weather. Honeydew are like watermelons, needing about 100 days to ripen. Cantaloupes tend to mature faster and need 75 to 85 days, depending on the variety.

Melons are traditionally grown on hills because they need fertile, well-drained soil to produce well. It is easy to build up a 3-foot hill of rich compost, instead of heavily fertilizing the entire garden space. The raised hill also aids in good drainage.

The Boule d'Or melon is an example of an heirloom melon that isn't available commercially, but is fabulous in the home garden.
(Photo courtesy of Baker Creek Seeds)

You can sow melons directly into the prepared garden area, or start seedlings indoors three weeks early. If you are growing seedlings indoors, keep the light source close to the seeds to discourage leggy growth. You'll also want to be careful to harden off the seedlings so they don't get stressed from the shock of cooler temperatures. Tradition dictates growing three plants per hill and spacing them 6 feet apart, but I prefer to grow one plant per hill and space them about 3 feet apart. Growing one plant per hill lessens the competition for water, which is always a premium in our area.

Heat stress will hurt the production of your melon vines, as well as alter the flavor of the melons. The more leaves, and the healthier the leaves, the sweeter and more nutritious the fruit will be. If the fruits will be smaller, you can grow them up a trellis. For larger melons put a board underneath to keep them from resting directly on the ground.

Muskmelons and honeydew will slip easily off the vine when they are ripe. Watermelons will show a browning of the curled tendril nearest the melon, although it's fun to "thump" them and impress your friends with your watermelon mind-power. Some cultivars change colors when they are ripe, making it even easier to tell.

Melons can be rather pest-prone with wilting diseases and powdery mildew topping the list. To prevent aphids and cucumber beetles from spreading disease to your plants, try covering growing vines with a row cover until the flowers appear and need to be fertilized. Squash vine borer can also infect melons, so watch for the signs.

Altaiskaya. Bushy vines that only grow 4 to 5 feet long and produce 3-pound fruits. Does well in shorter seasons as they mature within 75 days. Suitable for container growth as well.

Green Flesh. A sweet honeydew with small seed space and plentiful bright green flesh. Matures in 115 days at 3 to 6 pounds.

Jenny Lind. This popular cantaloupe matures in 80 days and weighs between 1 and 2 pounds. A sweet melon on prolific vines.

Moon and Stars. An heirloom watermelon that is regaining popularity in home and specialty gardens. A very sweet melon that has dark green skin with yellow splotches, and the fruit matures at 10 pounds.

Sugar Baby. Watermelon that matures in just over 80 days. This fast-growing melon vine produces large, sweet fruit.

Strawberries (*Fragaria spp.*)

Strawberries are herbaceous perennials that produce delicious red (or occasionally white) berries. They make useful ground covers in full-sun areas and can be interplanted between or in front of larger shrubs and plants. Most strawberries grow between 6 to 12 inches tall and wide. Alpine strawberries stay in neat mounds but other strawberries can spread through running stems.

This strawberry jar is an easy way to grow strawberries even with little room. It will produce fruit the first year if you start with strawberry crowns.

Strawberries need at least four hours of sun and prefer well-drained soil that has lots of organic matter or compost mixed in. Mulch will help keep strawberries from being crowded out by weeds, and preserve moisture in drought-prone areas where strawberries won't thrive.

Some strawberry types produce a lot of berries all at once, while others produce throughout the year. The former are the kind I like best for landscaping purposes so the pretty white flowers will appear here and there all season long. Unlike many other fruits, you can harvest strawberries from the very start. Harvest your berries as soon as the color turns, but watch out for marauding birds. A covering of bird netting can help prevent loss if it becomes a big problem.

Strawberries don't store in the fridge for very long but you can freeze them, bake them, or create jams and preserves to store them.

> **OVER THE GARDEN FENCE**
>
> To store strawberries as long as possible, they need to be dry and unwashed. Wrap them in a paper towel so they aren't touching each other and put them in a lidded glass jar in the fridge. This way they will store for about a week. Wash just before using them.

Perennial Edibles and Nuts to Grow

These perennial plants, shrubs, or nut trees make great fruiting additions if you have the space to dedicate to them. Some, like asparagus, are perennials that need plenty of space. Others, like walnut, are fabulous shade trees but can inhibit the growth of other plants in the area.

Asparagus (*Asparagus officinalis*)

Asparagus is a perennial vegetable that is rather expensive to purchase in the grocery store, but very easy to grow for yourself in a garden. And if you are the type of person who likes to maximize his time, energy, and money, you'll be pleased to find out that asparagus is a perennial vegetable. This means you will only have to plant asparagus once, but you'll have a harvest year after year.

Asparagus plants are often available as 2-year-old crowns with a long, stable root system already in place. Because asparagus is a perennial vegetable, I have had the most success purchasing the plants already started in this fashion. Asparagus is available in seed, but it will take two or three years before any shoots of harvestable size are produced. By planting 1- or 2-year-old crowns and roots, you can begin to harvest the very next year.

These perennials are what many call a "heavy feeder." This means that it needs to have plenty of rich, organic material to keep the soil well fertilized. This can be as simple as a thick layer of mulch or compost added onto the garden bed each year. Maintenance is easy with a yearly trimming of old fronds in the late winter or early spring. This will allow room for the new shoots that appear in early spring.

Harvesting asparagus plants couldn't be any easier. Stalks that are thicker than your pinky finger can be snapped off at soil level, rinsed, and then cooked in whatever way you most enjoy. Stalks that are smaller than your pinky finger should be allowed to grow into leafy fronds to help the plant gather energy and nutrients. They will be larger the following year and your harvest will increase. Usually your home-grown spears will be 6 to 10 inches long. Gardeners should watch their asparagus closely in the spring, and harvest spears before the tips begin to open.

These asparagus spears can be grown year after year for very little time or money investment. A bundle like this purchased from the grocery store is not cheap.
(Photo courtesy of Rachel Matthews)

Asparagus is such a long-lasting plant, a few dozen roots will provide produce for a family for many years. The taste of the home-grown spears versus store-bought spears is incomparable, and the savings of growing your own is significant.

Elderberries (*Sambucus canadensis*)

Elderberry is a wild shrub in our area and grows almost anywhere along the unmowed roadsides in the summer. We dug up a few starts from a friend's house that were taking over their fence line and transplanted them to the ranch.

Elderberries are not typically eaten fresh, but more often used in jams, wine, and tarts. We also use elderberry syrup for its therapeutic benefits to help minimize flu and respiratory illness symptoms.

The shrubby growth tends to be pretty loose and open. Elderberry is hardy in zones 4 through 7 depending on the cultivar, and when grown in favorable conditions it becomes a large, sprawling bush.

Elderberrries are pretty easy to grow and tolerate drought better than many other fruits. The flowers (also edible) are large white umbrels, resembling Queen Anne's lace but much larger, often as big in diameter as a dinner platter. The fragrant white flowers give way to huge clusters of dark purple, nearly black berries. And while the berries can never be said to be sweet, they do have a unique flavor particularly well suited to wine and jam.

These sprawling elderberry bushes have large clusters of flowers that are already starting to turn to berries in a few places on the bush.
(Photo courtesy of Tatiana Gerus)

Harvest berry clusters when they are ripe by simply cutting them off at the stem. You can easily use cheesecloth to strain out the seeds and leave the juice for syrups, wines, and jams. Watch out for birds that will literally flock to your harvest and may beat you to it if you aren't careful.

Pruning is easily done by simply cutting back the oldest canes each year, leaving half a dozen of the new, strong-looking stems. An established elderberry bush can even survive being accidentally mowed down with the brush hog.

Named cultivars have been developed for tolerance to various regional climates, as well as productivity, flavor, or size of the berry clusters. For example, Black Beauty does best in cooler climates and has pink flower blooms, while Wyldewood has huge berry clusters and ripens later in the season making it better suited to southern growing areas.

Mulch newly planted shrubs to smother out weeds until the plants are well established. Trouble free, disease resistant, and drought tolerant, elderberry is a great addition to the landscape if you're able to give them room to spread.

Pecans (*Carya illinoinenmis*)

Pecans have various hardiness depending on the cultivar. Also note that these gorgeous trees have varying disease resistances as well. Pecan trees are lovely shade trees and I love that they have the added benefit of a yearly crop of delicious nuts.

Unfortunately, even the hardiest cultivars won't usually produce reliably beyond zone 5 as pecans need a longer season for the nuts to mature. But oh, what a delicious harvest you can obtain if you live in zones 5 through 9.

This newly planted pecan will quickly develop a large taproot and in a few years become a lovely, productive tree.

Starting pecan saplings from trees that are just a few years old can take some time. While pecan trees are fairly drought tolerant when they are well established, they need extra water when getting started. Pecans have been developed commercially for so long now, there are named cultivars suitable for most regions within the growing zone. A quick check with the extension office will help you narrow your options.

You'll want to have more than one pecan tree in the area for the best pollination. While pecans have both male and female flowers, sometimes the male pollen is produced before the female flowers. By having two different types of pecan trees within 100 feet, you'll be sure that more pollen is available when your tree's female flowers are receptive.

Pruning is as simple as removing suckers from the roots, and any side branches that are growing below your clearance tolerance. Our pecan tree would become so heavy-laden with fruit that branches below the 10-foot clearance line would literally touch the ground.

Harvesting is rather simple as well, because nuts fall to the ground when they are ripe. But be quick—send your kids out to pick them up before the squirrels eat them all!

Walnuts (*Juglans spp.*)

Different types of walnuts are available, and you can find some hardy and productive ones in zones 4 though 9. The butternut (*J. cinerea*) is a walnut relative, sometimes called the white walnut, which is even hardy to zone 3. Walnuts make a fabulous shade tree as long as you have room in the landscape for a 50-foot-tall tree.

THORNY MATTERS

Black walnut can inhibit the growth of certain other plants in the landscape, so plant it in an otherwise unused corner of the yard, or plant tolerant plants nearby.

Walnuts prefer slightly alkaline soil and are usually best planted from seed. That's because the young trees quickly develop a thick taproot that acts to stabilize the tree, and provide moisture during dry seasons. On the downside, that means you'll have a few years of waiting before a newly planted tree will produce edible nuts for you.

The alternative is to plant a grafted walnut where a walnut-producing top has been grafted onto the rootstock of another tree species. This makes the tree more tolerant to being transplanted and can cut a few years off your wait time.

As with pecans, you should prune away suckers that develop from the root system, and any branches that die during winter or summer seasonal extremes. Otherwise, pruning needs for walnuts are minimal and a tree will usually grow well on its own.

When ripe, walnuts fall from the tree on their own. A good shake of the lower branches might help speed things along. Otherwise check the ground below the tree every day while nuts are ripening, every other day at the most, so the nuts don't rot on the ground.

Let the nuts dry out before shelling so the meat of the nut will be more easily accessed. Nuts will store for a long time in the freezer if you have an abundance. With the high price of nuts purchased from the grocery store, a mature nut tree is an invaluable asset to the backyard farm.

Nuts turn rancid when they are exposed to oxygen. If you don't have room in the freezer to store nuts, you can store them in a glass jar with oxygen absorbers and desiccants (available from storage supply shops) to prevent rancidity.

Heirloom Plants and Saving Seeds  11

What does gardening with heirlooms mean? You've probably heard the term before, because there is a growing resurgence of heirloom gardening. And for good reason.

While definitions vary, most gardeners generally consider heirloom varieties as 50 to 100 years old. The other key factor of heirloom plants is that they are open-pollinated. This means the seed breeds true. For example, the seeds of an heirloom tomato will produce new tomato plants that look just like the parent. Hybrids, on the other hand, are mixed of two different parents and so won't breed true.

Benefits of Heirloom Gardening

Heirloom varieties developed from families in various regions saving the best plants each year for seed planting the following year. Whereas many gardeners today buy plants already started (at a much higher cost, I might add), it was the norm in years past to save seeds for each new year's planting. Why are people rediscovering this trend? Because there are a lot of benefits in doing so.

Cost Effectiveness for Growing Produce

It is just a touch more work to save seed year to year and start the majority of your vegetables from seed, but there are so many benefits for the backyard farmer. One of the biggest benefits is the cost savings! Last year my father-in-law planted purple-hulled peas to grind for feed (and to feed us) for the cattle using seeds that were passed down from his father.

If you are planting even a couple dozen tomatoes, you will pay nearly $50 just for the transplants. But if you purchase three packets of seeds (for three varieties of tomatoes), you will spend $6. That's a savings of about $40 for the year. Fast-forward to next year. When you save the seed from the open-pollinated, heirloom varieties, your price for the tomatoes you plant the following year is $0, whereas you would spend another $50 to plant tomato starts. And the next year? And the next? Multiply this effect in a larger garden area, with multiple crops per year, and you can see how this will really add up! Even if you save and use seeds for half of your produce, you can see how this will give you a tremendous savings on your backyard farm.

Only seeds from open-pollinated varieties will produce the same plant as the one from which the seeds were harvested.
(Photo courtesy of Broadfork Farm)

Sustainability and Self-Sufficiency

Sustainability is an important aspect of the backyard farm. You want to be able to keep things going year to year without having to put more into the farm than you are getting out. After you've made that initial investment, you want to be able to have a yearly return with minimal reinvestments. This is true of your chickens, bees, goats, and most certainly true of your plants as well.

Of course, some of it is the control factor. After all, seeds are food. And at the end of the day, if you have food, you have the ability to live. With so many of our foods now being regularly imported, or grown from seed that produces sterile offspring (in other words, the plants grow one year and no more), the food situation in America is more tenuous than most people realize. So taking a little bit of control back for yourself is not only a cost-saving measure, but also a little step toward independence.

Hybrids are often more readily available in seed catalogs for a number of reasons. Seed companies can keep the parentage of hybrid varieties a secret, making them easier to sell at a higher price. And because hybrids will not breed true from seed, a seed company knows that gardeners will have to come back and buy hybrid seeds from them again next year. And the next year. By making the switch to open-pollinated varieties you can break that dependence.

The other very cool aspect of sustainability that comes into play with heirloom varieties is the ability to select for regional tolerances. For example, there is a tomato variety that originated from Arkansas called the Arkansas Traveler and it is known for its heat and drought tolerance. Meanwhile, Caspian Pink, which originated in Russia, is better suited for cooler climates. These regional variances can actually help a backyard farmer who is planting heirloom varieties better suited to their specific growing conditions.

So many varieties of plants are available via seeds as compared to commercially available transplants. For the adventurous gardener, starting seeds is a must!

These tolerances developed naturally as each year homesteaders would select the best, healthiest, tastiest, and most productive plants as the parent plants for next year's seeds. Plants that were less tolerant of local growing conditions, pests, and diseases were eliminated as seed plants. So those weaker genes were bred out of the variety. This makes it easier for gardeners who want to grow their gardens using Earth-friendly, organic methods to do so.

One of the biggest concerns with the decline of heirloom breeds in the past 50 years is that of genetic diversity. If all corn was exactly the same, it would be more easily threatened by a disease outbreak or widespread pest infestation. But if there are 1,000 different types of corn, a disease outbreak might kill off a few strains of corn, but it's likely that others would have a natural immunity or resistance. This genetic diversity has decreased in the last few decades so some gardeners are now working hard to preserve heirloom varieties that might otherwise die out.

It can be a challenge for gardeners with a small space to save seed from a large-enough number of plants to maintain a diverse genetic makeup. That's why Seed Savers Exchange (seedsavers.org) and other organizations can come in handy! Every couple of years consider trading seeds with someone from another area of the country to freshen the genetic pool you're working with. This is especially important if you are saving seeds from just a couple plants each year. If you have more space, and are able to save seeds from a dozen or more plants, that might not be necessary.

OVER THE GARDEN FENCE

Seed Savers Exchange is a seed-saving organization dedicated to saving and sharing heirloom seed varieties. Get their catalog to see the amazing variety of plants available. There are also seed swap clubs and organizations that allow you to trade your seed varieties for new varieties from other gardeners.

How to Save Seeds

Some of the easiest vegetables to start as a seed saver are beans, cucumbers, peas, peppers, squash, and tomatoes. There are two main ways to save seed. While the details of each type of plant you might grow in the backyard farm is beyond the scope of this book, I'll discuss the two primary methods of seed cleaning.

Wet processing involves removing seeds from fruit while the fruit is still wet, and then there's usually a fermentation process and a drying process. *Dry processing* involves allowing the fruits to dry up on the vine and then winnowing the seeds from the plants. Some common fruits and vegetables that are processed with wet processing include melons, squashes, and tomatoes. Common fruits and vegetables where you'll use dry processing include beans, peas, and lettuce.

Processing or fermenting wet seeds sounds complicated, putting some gardeners off the attempt, but it's actually easier than you might think if you'll give it a try. My husband's grandmother used to toss her tomatoes onto the pile of muck outside the chicken coop in a sheltered location. And each spring she'd have seedlings to transplant into the garden. I'm going to share a slightly surer method than "toss your tomatoes in the compost heap over winter" but hopefully you realize it's not too difficult.

Large fruits like melons and pumpkins can be cut open and the seeds and pulp scooped out. Smaller fruits like cherry tomatoes can be smooshed up. Often these will need to go through a fermentation process that mimics what the fruit would go through falling to the ground to rot. If you take the seeds through a fermenting process, you won't want to add a lot of water to the mixture as that can slow down the process.

Tomatoes are a great place to start with seed saving because most tomatoes are self-pollinating. This means that each flower pollinates itself and you won't have accidental crosses even when you grow different types of tomatoes in the same garden. So I'll use tomatoes as an example of how to ferment and save seeds.

Squeeze all the mush from the inside of the tomatoes you're using as seed into a yogurt cup or other disposable container. If you're saving more than one variety, be sure to use a different container for each and to mark them so you can tell which seeds are which! Add in just a touch of water, not more than a cup, to help the seeds separate. Mix up the mixture to help stir it all together.

Squeeze the pulp into a jar or container.
(Photo courtesy of Baker Creek Seeds)

Let the tomato mixture sit out at room temperature (or warmer), fully exposed to the air. If you have an outdoor shed or garage, you'll probably want to put the mix out there because it will stink. Whatever you do, don't place it within reach of your kiddos who might knock it over into a floor rug where it will cause a huge stench in the house.

The mixture will grow a mold over the top of it. This mold will break down the gel material that surrounds the seeds and prohibit germination. This process also helps destroy any illnesses or diseases that might be lingering from your garden. The fermentation process can take one to four days and you'll know it's finished when the mold layer covers the top of the container completely.

This fermentation cycle is complete as the mold has completely covered the mixture.
(Photo courtesy of Baker Creek Seeds)

Now add extra water to the container so it's at least doubled in volume. If you stir everything up a bit you'll see the plump, fertile seeds settle to the bottom of the container, while the unusable seeds and mold will float at the top. This way you can pour off the mold and unusable seeds and still keep the good seeds in the container.

The moldy mixture has had fresh water added to it so the mold and unviable seeds can be poured off.
(Photo courtesy of Baker Creek Seeds)

Pour the rest of the good seeds into a strainer and rinse them clean. When just the seeds are left in the strainer, place them onto a saucer or small plate to dry. Again, be sure you keep the seeds from each variety separate and marked.

These tomato seeds have just finished the fermentation process and are ready to dry and store for next year's growing season.
(Photo courtesy of Baker Creek Seeds)

Let the seeds dry in a place where they are out of direct sunlight and high heat, and be sure to stir them up a bit each day to prevent molding or clumping. When they are completely dry you can store them in a cool, dry, and airtight location where they will remain viable for 5 to 10 years.

Here's a great tip from Jodi of prepperkitchen.com: When drying small seeds like tomato seeds, lay them on a paper towel over a screen to finish drying. They will dry and stick to the paper towel, making it easy to roll up the paper and store the seeds when they are completely dry. The next spring you can easily plant the seeds by ripping off pieces of the paper towel with the seeds stuck to it, which makes it easy to space the seeds without overcrowding them. The paper towel simply decomposes while the seed grows into the new plant. Brilliant!

Dry processing involves allowing the seeds to dry directly on the plant, usually inside a pod of some kind. After the plant and seeds are dry, they can be harvested and threshed from the plant. With plants like sunflowers, I cut the entire flower head and hang it upside down in the laundry room to finish drying. With other plants, like beans, or the purple-hulled peas we grow at the ranch, the pods are allowed to dry on the plant. Then the pods are harvested and the seeds removed from inside the pod.

Lay the seeds out to make sure they are fully dry. Keep them in a well-ventilated area out of direct sunlight until they are completely dried. You'll know a bean or pea is dry when you can strike it with a hammer and it shatters instead of smashing. Store in a cool location and your seeds should be viable for a couple years.

THORNY MATTERS

Weevils can destroy your seed crop if you store your own bean seeds. These worms can live inside a bean undetected until you store the infected bean in a jar with uninfected beans. By next year you'll have no viable seeds left and you never even knew you were under attack! If weevils are a problem in your area, place your seeds in the freezer for at least three days before storing them to kill off any weevils. Make sure the beans are fully dry before freezing them.

Testing Viability and Starting Seeds

Growing vegetables from seed can be a big step toward a self-sufficient backyard farm. And, as I mentioned, it's also a huge leap forward in terms of the financial benefits. Some plants are so easy to grow from seed that it doesn't make sense to buy them as more expensive starts.

You can test the seeds you're saving or starting to see if they are *viable,* or able to germinate and grow into a plant. If you only have a small amount of seeds, you can test with 10 seeds. Commercial lots are tested with 100 seeds to give a truer sample.

Most seeds can be tested through simply placing the selected number of seeds on a damp paper towel. Cover the seeds with another damp paper towel and place them rolled up gently into a loose bag. You want the seeds to stay damp, but not smothered—they need to breathe. Place the seeds in a consistently warm area that stays around 75°F, depending on the specific plant you're testing. Check your seeds every day to see how many have sprouted and rewet the paper towels if necessary. As the seeds sprout, you can remove them and after two weeks you'll have a good estimate of the percentage of seeds that are viable. For example, 7 out of 10 would be a 70 percent germination rate.

Starting seeds begins with good-quality seed that has been stored well. Moisture and warmth can trigger germination or rotting so your seed should be kept dry, cool, and out of the light until you're ready to plant. Many seeds can be started indoors, and then the plants transplanted outdoors when the weather is right. Other plants do best when the seeds are sown directly outdoors.

Seeds are like beach tourists—they tend to come out only when the temperature reaches 75°F or so. That's why it's sometimes easier to start certain plants indoors. As a general rule, unless it's a plant such as spinach, lettuce, or another fall-season veggie, you'll want to provide warmth for your seeds to help them germinate.

No matter how you start your seeds, there are a few things that all seeds need to grow well and develop into productive plants.

These seedlings are just beginning to sprout in the warmth of the sunny window and grow light.

Temperature. Seeds need to reach the correct temperature in order to germinate. Until that minimum temperature requirement is met, they stay dormant. It's this dormancy that allows the seed to store for so long.

Some indoor seed starting kits have heating pads under the container to warm the soil. Other gardeners use cold frames, greenhouses, or warm windowsills to start seeds ahead of the season. Grow lights can provide both light and warmth for growing seeds.

Light. Seeds also need light to grow. Whether it's sunlight from a window or an artificial light doesn't matter. Because sunlight levels are so vital for a healthy garden, it's no surprise that light is important for the plants' seeds as well. The most important aspect of light is not how long it lasts, but how bright it is. If light is too weak, the seedlings will grow long and lanky, and be much weaker when transplanted to the garden.

If you grow your seedlings in a window so they'll have natural light, instead of grow lights, be sure you turn the seed tray each day. Seedlings will naturally grow toward the light, so if you never turn the tray your seedlings will have a noticeable lean.

It's not hard to rig a grow light to be just a few inches above your seeds (within 3 to 6 inches is ideal).

Soil or seed mix. Let the debate begin! Peat moss? Potting soil? Organic-and-specially-created-and-thus-overpriced-seedling soil? What on Earth should you plant your seeds in anyway? Some of the seed starting kits come with little transplantable pots that can help make that step easier for beginners. Other gardeners love to come up with their own seed starting mixes—especially when there are a lot of seeds to be started. When you move beyond a single kit, more than a couple dozen plants or so, it will probably be more economical for you to start seeds with your own soil.

Most seed starting mixes come out to something pretty close to one third organic matter (compost or peat moss, for example), one third moisture retention and drainage matter (perlite or vermiculite), and one third soil. Oftentimes gardeners will use a screen to sift out finest soil particles and only use that in their seed starting mixes to make it easier for the seedlings. Most commercial seed starting mixes will have something close to this type of blend.

Moisture. As with all plants, seedlings need water. Very small seeds can be washed away by heavy watering, so water with care! Some gardeners use a mister or even a simple spray bottle set to mist setting. Other times seed starting kits are watered from below, eliminating the need to disturb seeds at the soil level at all. Whatever you use, keep the soil consistently moist but not soggy.

Growing your plants from seed adds another level to your vegetable garden planning, and saving your own seeds for the next year is another level yet. Be sure to check out the Seed Starter Log in Appendix B's journal pages.

Animals for a Backyard Farm  3

This part explores the animals of the backyard farm, with a special focus on those that are well-suited to live comfortably in a small-space situation. The emphasis of care is always on the least-invasive, most comfortable, and respectful of the natural impulses of the animals as possible.

Organic and sustainable care is discussed for chickens, rabbits, sheep and goats, and bees. Each chapter covers basic husbandry, times when special care is required, and the reasons for choosing to work with an animal's basic nature instead of against the basic instincts.

Animals cared for in a well-maintained system will not suffer the diseases of overcrowding, poor eating, stress, and misuse as animals in the typical commercial systems. Learn how to create a system that respects the land and your animals at the same time. Whether you want to raise animals for meat, eggs, milk, fiber, or honey, you'll find some great options for getting started.

Keeping Chickens on a Small Scale  12

Backyard farmers who want to be self-sufficient will need a good source of protein. And one of the most complete sources of protein available is found in the egg. And on the backyard farm, what an egg you will receive!

The fresh eggs you'll collect from your backyard chicken flock will be nothing like the average grocery store-purchased egg. Pasture-raised chickens (this is what you'll be raising) produce eggs that are far more nutritious (not to mention tasty) than commercially produced eggs. Research conducted by *Mother Earth News* compared the nutrients of sample eggs with the USDA nutritional information of the average eggs. What they found is that pasture-raised eggs have a third less cholesterol, twice the levels of omega-3 fatty acids, at least 50 percent more vitamin A, and higher levels of folic acid and vitamin B_{12}.

You'll notice that these eggs look different as well. Pasture-raised eggs have up to seven times the levels of beta-carotene, which gives them a noticeably more yellow-orange color. The yolks are stronger and firmer, and even the egg whites are less watery.

And that same comparison applies to the meat of pasture-raised chickens. A study done in 1999 at Pennsylvania State University tested the nutritional content of the meat and found that pasture-raised chicken has 20 percent less fat, 30 percent less saturated fat, and twice as much vitamin A compared to standard, commercially produced chicken.

This chapter covers all the basic information you'll need to get started raising your own flock of backyard chickens. Even in a small space you will be able to raise enough chickens for fresh eggs for your family. And if you choose, and have just a touch extra time and space to devote, you can raise extra chickens specifically for meat production.

Breeds for Backyard Flocks

Some chicken breeds are better suited to a backyard flock than others because the traits that are bred out of commercial egg-producing chickens are actually traits that might come in handy in a backyard situation. *Broodiness* means fewer eggs as that hen will stop laying, but it is a free means of replenishing your laying stock, so there are pros and cons depending on your desires. Winter hardiness isn't needed in a flock that lives in climate-controlled situations for its short life, but for a backyard flock it might be a positive trait.

DEFINITION

A **broody** hen is one that "goes setting" or sits on her eggs to hatch them. A hen in brood should be separated from the rest of the flock (if you wish to allow the hen to hatch her eggs) and placed in a small pen with food and water close to her. She will stay on the eggs until the eggs hatch. Broodiness is triggered by the longer days in the spring and is often appreciated on the backyard farm, where free baby chicks are considered a plus. This trait has been bred out of most commercial laying strains because they prefer the hens not take a break from laying eggs.

This hen is displaying the classic broody posture. Notice how she's laying flat against the ground, spread out to cover a wide area with her warmth, and has her feathers fluffed out to increase warmth and insulation for her babies.
(Photo courtesy of Sammydavisdog)

The following breeds are just a few of the many hundreds of breeds that are available. These are some of the most popular and useful on backyard farms and several of them are endangered heritage breeds. According to the American Livestock Breeds Conservancy (ALBC), a heritage chicken breed must be able to reproduce naturally; should represent a standard, recognized chicken breed; and must have a slow growth rate and long life span when raised in a traditional backyard environment.

Egg Layers

These breeds are those that are bred specifically for laying eggs and will produce more than 200 eggs each year. They are generally not considered useful as meat birds because none of these breeds mature to more than 5 pounds.

Ancona. Ancona is a hardy breed that lays between 220 and 240 eggs each year. The Italian breed is a very attractive breed that is black with white speckles. While they do tend to be more flighty, they can be tamed and their activity level makes them good foragers on the homestead. They are consistent layers that do well through the winters, but their large combs can freeze in extreme temperatures. Their small size makes them less desirable as a meat bird because they only grow $4\frac{1}{2}$ to 6 pounds but can also be broody, making them a nice addition to the flock for possible hatchings.

Hamburg. The Hamburg is a small but highly active breed, and comes in a myriad of attractive color varieties. They are better suited to foraging on their own as their small, quick nature makes them harder to catch in a more rural situation with lots of potential predators. This tough-to-catch bird is noisy and not for those who want a chicken that will be more of a pet. The glossy white eggs are larger than a bantam egg but the birds aren't much bigger. Their rose comb makes them more cold-hardy and less prone to frostbitten combs. You should get at least 230 eggs per year.

Leghorns. Leghorns are traditionally thought of as white; however, this breed is available in white, brown, buff, and rarely in blues and barred. All Leghorns produce white eggs. They are the premiere egg layers and can lay up to 300 eggs per year. Leghorns usually do not grow very big and aren't used as meat birds because they weigh only about 5 pounds at maturity. Even though they have a reputation for being nervous, Leghorns have, in my experience, been excellent foragers in the backyard pasture.

Penedescena. A Spanish breed that almost went extinct, Penedescenas lay the darkest brown eggs. They are still rare in the United States but four color strains are available: creole, black, wheaten, and partridge. They only grow to about 4 pounds at mature size; however, they are adaptable, attractive, and heat-hardy. Their unusual comb shape helps them tolerate the heat because the single comb becomes multilobed in the back, dispersing more heat. They lay approximately 200 eggs per year.

Meat Breeds

Cornish Cross. The most popular meat birds in the United States right now, these birds are bred specifically to reach full size in an extremely short amount of time. In fact, they are not usually considered suitable for long-term life on the homestead because they can outgrow their own bodies as the rapid weight gain can't be supported by weak legs and internal boney structures. The downside of Cornish Cross breeds is that they are not sustainable—you cannot raise your own to laying age and hatch your own for future generations. For inexpensive food-to-meat conversion, the Cornish Cross is hard to beat, though. They quickly grow to butchering size and are one of the fastest ways I know to fill your freezer with meat that you can control the quality of.

The baby Cornish Cross chicks are kept in a separate brooding pen. I call them "The Chicks Who Shall Not be Named" as it is always wise to avoid naming your future food.

Jersey Giant. Jersey Giants are the largest of the pure-breed chickens and mature at 10 to 13 pounds. They do lay a fair number of very large, medium-brown eggs but are generally considered a meat breed by virtue of their large size. Jersey Giants are slow growing, though, so aren't used in any commercial meat productions. In the backyard farm situation, they are excellent foragers, easy to handle, and known to be extremely cold-hardy. Hens are broody and make good mothers for their chicks, but you should be aware that Jersey Giant chicks can take 24 to 48 hours longer to hatch than other breeds.

Old Cornish. The Dark Cornish or Old Cornish are the other parent breed of the Cornish Cross and have a similar large size. Cornish grows 7 to 9 pounds but instead of maturing to butcher weight in 8 to 12 weeks, you will see harvestable sizes in 20 to 25 weeks. On the positive side, the flavor is said to be much better and their temperament is very calm. These birds build up size, and then bulk up layers of meat, instead of building up muscle mass at the risk of ultimate health. In this way they are a better choice for free range on the backyard farm.

OVER THE GARDEN FENCE

I was skeptical at the thought of raising my own meat when we first started that part of our self-sufficiency journey. However, the more I learned about the questionable practices of many commercial chicken farms and the health benefits of pasture-raised, organic chickens, the easier it was for me to accept the idea. Now that I've tried it and understand how much this hands-on process respects the animals we raise, I could never go back. It's well worth trying to see if you, too, get hooked on providing the healthiest possible meat from the happiest possible animals.

Dual-Purpose Breeds

Dual-purpose breeds are just that—the breeds of chicken that can do more than one thing well. While most of the breeds don't lay as many eggs as the egg layers, and they might not grow as quickly as the meat breeds, these dual-purpose breeds can usually do both things adequately. They often have other qualities that can be of benefit on the homestead such as good foraging skills, calm temperaments, or acting as good mothers able to rear their own offspring.

Ameraucana. One of my favorites, I love these friendly and docile birds that are known for laying blue eggs. Often confused with Araucana chickens, the Ameraucanas have ear muffs and tail feathers. (Araucana chickens, which also lay blue eggs, have ear tufts and no tail feathers.) They are good egg producers, laying three eggs a week average, and grow to a medium size of 6 to 7 pounds. Because they are bred for feathered ornaments of ear tufts, and for the blue egg–laying gene, Ameraucanas come in eight varieties of colors including wheaten, white, black, blue, silver, and brown.

One of my Ameraucana hens with her endearing and comical ear muffs.

Araucana. Araucana chickens are another blue egg–laying breed and they have ear tufts but no tails (called rumpless). Araucanas don't lay as many eggs as other breeds (about three a week on average) and end up slightly smaller at about 5 pounds. They are very good brooders and their pea combs make them cold-hardy. Even though they will be out-laid by your other hens in the summer, your Araucanas may well be the best layers in the winter months. They are adaptable birds that do well in both confinement and free run.

Minorca. Minorca chickens are one of the larger white-egg layers and can reach 6 to 7 pounds in size at maturity. They lay 220 to 240 very large white eggs when they fully mature. Some strains are reported to be more flighty, while others are friendlier. Minorcas are usually glossy black but there are buff strains and white strains as well. They have large combs and wattles that are subject to freezing in cold winter areas, but they are more tolerant of heat and humidity than other breeds might be.

Welsummer. This rare breed produces large, red-brown eggs at the rate of about 220 eggs per year. Welsummers are a pretty black and brownish-red color, and they mature to an average size of 6 to 7 pounds. The birds are friendly and even the roosters stay nonaggressive, which is great in a small backyard where the chickens will be running loose. Welsummers are considered good foragers in pasture and grassy areas, and will also be occasionally broody.

Orpington. Round and fluffy in appearance, Orpingtons are actually larger than you might think and mature at 8 to 10 pounds in size. My favorite variety is the Buff Orpington; however, black, blue, and white varieties are also common. The iridescent sheen of the orangey buff coloration is a delight to watch in the backyard. Moderate egg layers, Orpingtons produce three to four eggs per week, and tend to lay well throughout the year because their large size makes them more tolerant of winter weather. Their heavy stature makes them poor fliers and allows them to be contained with only casual fencing.

One of my Buff Orpington females, already feathered out and moved to her outdoor yard, but not laying yet.

Black Australorp. While Australorps are generally moderate egg layers, producing three or four eggs per week, a Black Australorp hen holds the record for laying 364 eggs in a single year. Australorps mature at more than 6½ to 8½ pounds and produce good meat. Even though the feathers are a dark, glossy black, the skin is white. They were originally called "Black Orpingtons" and share the same friendly demeanor and good parenting skills as the Orpingtons. Good year-round egg production seals the deal for many backyard flock owners, making Australorps a popular and logical choice.

Plymouth Rock. Rocks are a large, dual-purpose bird that were once the primary meat bird in America. They mature at 7 to 9 pounds and are fairly quick growing in the backyard flock. The hens lay brown eggs and can be expected to produce several eggs each week. The chickens are very cold-hardy, and they tend to have a calm and nonaggressive temperament. One thing to be aware of is that strains of the White Rock are the parent strains for Cornish Cross broiler hybrids, and these strains may be unsuited for living in a backyard flock. Choose your Plymouth Rocks from breeders that will provide you with the heritage strains that are true to their American origins: economical, intelligent, and friendly. Barred Rocks, the black and white feathered variety, is one of the most popular varieties and is highly attractive.

Rhode Island Red. One of only two chicken breeds that became a state bird (the other is the Jersey Giant), the Rhode Island Red is a beautiful breed with roosters in the iconic rooster coloring. Rhode Island Reds are prolific layers that produce more than five eggs per week in the first year. They mature between 6½ and 8½ pounds relatively quickly and are one of the parent breeds of the Black Sex Link layers. While the breed is generally considered friendly, roosters of certain strains can become aggressive with age and should be culled from the backyard flock.

SEX-LINKED HYBRIDS

Most chicks look the same in coloration in both male and female chicks. Sex-linked breeds and hybrids have different coloring in the males and females, making it easy to sex the chicks just by looking at them. For example, when a Rhode Island Red rooster is crossed with a Barred Rock hen, the female chicks are born black and mature black with reddish neck feathers, while the male chicks are born black with barred (white) markings and mature to display scattered white and barred markings throughout. Rarely are pure chicken breeds able to be sexed visually as chicks but those breeds include Cream Legbar, Amrock, and Barred Plymouth Rocks, among a handful of others.

Cuckoo Maran. This dual-purpose breed from France looks similar to the Barred Plymoth Rock but the barring is lighter and more muddied. The Cuckoo Marans lay a deep chocolatey-brown egg that is often speckled. Expect 150 to 200 eggs in a year and a mature size of 7 to 8½ pounds in the average Cuckoo Marans. Some strains rarely go broody, while other backyard farmers report their Cuckoo Marans do nothing but brood, so it seems to vary a lot depending on the breeding line.

Chantecler. The only truly Canadian breed, this endangered heritage breed was bred specifically for cold tolerance. The minimal comb and wattles prevent frostbite damage and the hens lay well through the winter months. Chantecler hens will lay around 120 to 180 eggs in a year and they will be large eggs of a medium-brown hue. There is a white variety, and a partridge variety that was bred to be better camouflaged while out foraging. The hens mature about 6½ pounds and roosters get to almost 9 pounds, making them a nice dual-purpose, cold-hardy breed for a backyard farmer who wants to help preserve a heritage breed.

Wyandotte. Bred in America in the late 1800s, the Wyandotte is a great dual-purpose breed that does well in the backyard farm situation. They also tolerate confinement better than more active breeds which makes them an excellent choice for a small-space flock in a city backyard. Wyandottes are laying machines, averaging around 200 eggs per year. The hens mature to just under 7 pounds and the roosters reach 8½ pounds. Wyandottes have an interesting silhouette with a deeply curved back and this, combined with the beautiful color variations, makes them a beautiful breed to look at. Wyandottes are most commonly available in silver-laced, Columbian, white, and golden-laced variations.

These are only a handful of the hundreds of breeds you would have access to for your backyard farm. You can see more about the breeds I didn't cover here at BackyardFarmingGuide.com/chickenbreeds.

Chicken Husbandry

Raising chickens on a backyard farm looks nothing like the commercial chicken houses. If you have a picture in your mind of dusty, smelly, noxious conditions from when you've driven past a rural chicken house, please realize that is an extreme example of overcrowded conditions and should never be replicated on your homestead. The goal of a backyard farm is to manage all the parts in a balance or harmony.

Chickens can be an easy-to-maintain part of the backyard farm. More than that, the chickens can be huge contributors to the health of the backyard farm. Ours are rock stars in the backyard and till our garden, control pests, feed the soil, and feed the family; they provide a great deal for us and have our utmost respect in return.

On our quarter-acre lot we house our small flock in a wooden chicken coop house and a portable chicken tractor or ark. This allows us to move the chickens around the yard (and keep them protected from dogs), while still giving them the freedom to forage for themselves. In the late fall we turn the chickens (and the goats) loose through the garden area where they can turn in any of the remaining plant materials, gobble up the grubs, and naturally fertilize the soil.

Chick Brood Box

When you first get your baby chicks from the feed store, mail order, or local rancher, you can house all your chicks in a plastic tote. This "brood area" can be as simple as a box in the back hallway, or as elaborate as a climate-controlled subsection to a large chicken house. For most backyard farmers trying to raise a few chicks, you can start the chicks in the house and move them outdoors when they have feathered out and can better tolerate changes in climate.

THORNY MATTERS

You may see pictures of chicks being started in cardboard boxes; however, I do not recommend it. I prefer to use a large plastic tote (which is easier to clean) or a metal dog crate out in a sheltered area of the shed. The cardboard box is also a fire hazard if the heat lamp over the box falls into the box.

For raising baby chicks you need the following:

- ⚘ **Brood box.** Whatever kind you choose should have walls that are tall enough to shield the chicks from any drafts.

- ⚘ **Heat lamp.** You need to provide heat for the chicks. The temperature should be within 90°F to 95°F.

- ⚘ **Chick feed and feeders.** The feeders with holes on top help keep the chicks from soiling the feed and wasting it.

- ⚘ **Water source.** I highly recommend getting an actual chick waterer (rather than using a water dish) to prevent accidental drownings or tipping over of the water dish. If chicks don't find the water dish on their own, you'll need to dip the tip of their beaks into the water.

- ⚘ **Bedding.** Pine chips, straw, shredded newspaper, or mulched hay can be used successfully. A flat, slippery surface can hurt the legs of your chicks, so make sure the bedding is at least an inch thick.

One of the easiest ways to monitor the temperature of your chick brood box is to simply leave the heat lamp hanging above it in such a way that you can raise it or lower it as needed. Place your heat lamp to one side of the box, not in the direct center, and then watch your chicks and adjust as needed. If the chicks are all huddled together in the corner under the lamp, they are cold and you should lower the heat lamp a little. If they are pushed to the far side away from the lamp, you need to raise the lamp as they are too hot. You'll know you have it right when the chicks are scattered about—some eating, some flopped over as if dead (don't worry, they are just sleeping), some here, some there.

These chicks display great behavior—some resting, some eating. They are happy and cheeky chicks.

Be sure not to overcrowd your brood box. You'll want to make sure that you have one feeder and one waterer for *every* 15 to 25 baby chicks. They need about 6 inches of space per chick when they are newly hatched, and by 8 to 10 weeks when they are feathering out, you'll want to provide at least a square foot of space per chick. Ours have moved outside by about 4 or 5 weeks of age so they do not become overcrowded in the house.

Chicken Coop

When your chicks are feathering out—a phase we call the punk-rock phase for reasons you'll discover—you can move them to their outside enclosure. We have one enclosure with smaller mesh that we move them to while they finish growing out. After they are adult size and may start laying eggs, we move them in with the rest of the flock.

There is no right way to build a chicken coop. You can build a chicken coop out of just about anything, but there are a few minimal requirements to keep in mind.

Space. You want to allow about 3 to 5 feet of space per chicken. Allow more if they don't have access to an outdoor run. You can get away with less room inside if they spend all their waking hours running amok in the yard foraging. A moveable pen will keep your chickens in fresh pasture even if you don't have a large amount of free-roaming space to work with.

Shelter. Your coop should provide adequate shelter from the elements, and this means both heat and cold. You shouldn't need to heat the chicken coop, if there are no drafts in the winter. Screened windows and doors, and placing the coop in the shade of a tree during summer months, can help prevent overheating if you're in a hot summer region like I am.

This small, sturdy chicken coop can be moved about the yard. Here, my son is pushing pieces of grass through the window screens.

Protection from predators. There's a country saying that if it has a digestive system, it will eat a chicken. Even chicken wire won't stop all predators, as we discovered when we erected a temporary pen of wire and had our flock decimated by a band of raccoons that chased the chickens to the edges of the wire, then reached through and slaughtered them all. Dogs, hawks, snakes, raccoons, opossums, and many other carnivores are potential predators of chickens.

In our area hawks are one of the biggest threats, so our chickens have mesh coverings across the top of their outdoor enclosure. If your chicken coop has a dirt floor you may need to bury wire a foot deep at the walls to prevent critters from digging underneath the walls and into the coop. Alternatively, you can place large cement or cinder blocks (or big rocks) that are too large for predators to move around the outside edges of your coop. Secure wire mesh screening on any windows and ventilation gaps or you may find a persistent stray will bust through.

Food and water. Even when our chickens are foraging free-range, we provide a higher-protein feed and oyster shell crumbles at all times. We also have fresh, clean water available for the chickens at all times and in the hottest summers we have waterers in the coop and out in the yard as well.

THORNY MATTERS

Chickens need fresh, clean water available at all times, especially as baby chicks. This means even if your chicken waterer isn't empty, you should be checking to see if the water is soiled. Be sure to keep the water clean or your chickens may get sick and even die.

Roosts and nest boxes. Providing your hens with a nice, enclosed area to lay their eggs in will not only make them feel more comfortable doing so, but will keep them from laying their eggs willy nilly. Nest boxes can be as simple as a 5-gallon bucket nailed sidewise on the wall with hay stuffed inside. Chicken roosts should be about 2 inches square, but rounded on top to make it easy for chickens to grasp. Provide just under a foot of roost space per chicken and keep them about 18 inches from the walls and ceiling. Consider giving them multiple roosts at a variety of levels so each chicken can find its preference and have plenty of room.

Chickens naturally perch on roosts. Providing them an outlet for their natural behavior will keep them happy.

Dustbox. Chickens clean themselves in two different ways—through preening and through dust bathing. A most comical sight, a chicken dust bathing will flap, flounder, practically roll over and bury herself, to throw dirt and sand up onto her back and into her feathers. The dirt absorbs oils and debris, which the chicken is then able to slough off with her preening. Dust bathing is also thought to help with mite prevention (discussed later in this chapter).

A good dust bath keeps your hens healthy. This Ameraucana is bathing and attracts the attention of the other hens in the flock.

Regular Maintenance

While chickens are a relatively low-maintenance livestock option, that doesn't mean no-maintenance. In order to provide everything they provide for you, there are certain minimum requirements that must be met. We teach our children that happy chickens make happy eggs.

Feeding your flock. There is a wide variety of feeding programs that range from 100 percent commercially purchased feed for a totally enclosed and dependent flock, to 100 percent free-roaming chickens that aren't supplemented with any feed at all. In some areas of Oklahoma I even see households where the chickens are practically feral and fend completely for themselves. Most backyard farms fall somewhere in the middle.

The less feed you have to purchase, the more self-sufficiency and the fewer feed bills you have. Feeding your flock directly from your kitchen or garden as much as possible just makes sense to increase their health and lower your overall expenses.

Chick starter has a high level of protein to help support developing chicks as they experience their rapid growth, so we always use this prepackaged food as the main source of food for our chicks.

When feeding baby chicks, we keep starter crumbles (the specially formulated food for baby chicks) available at all times, but also don't hesitate to provide fresh greens at an early stage. The chicks always seem to enjoy the diversion, and it helps supplement their food. These greens are considered a treat that we begin offering on a limited basis if chicks are healthy at week 3 or 4.

When your chickens are feathered out, you won't need to feed such high levels of protein. In fact, given enough room to run, your chickens will forage all their protein and most of their other feed requirements for themselves. We allow free-choice laying crumbles (higher-protein commercial feed for laying hens) for our hens and scatter some mixed grains to them in the evening.

My kids love to pluck seedy grass heads from the fence rows and push through the bars of their ark to watch them jump and squabble over them. Overripe melons, unsuspecting grasshoppers that allowed my boys to catch them, baked eggshells left over from our morning breakfasts, milk that got pushed to the back of the fridge too many days ago, and a myriad of other bits and scraps make their way to the chicken yard.

I know that my hens will be able to balance their own nutritional requirements with their commercially prepared feed always available. We find that often they go without touching it—especially in the summer months when more insects, produce, and green grasses are available.

Water. Fresh water should be available at all times. We have a large self-waterer and usually refill it completely when it gets down about halfway. Each day you have to check the water for any hay, scattered feed, or soiling that may have accidentally gotten into the water bowl. Raising it off the ground a little bit will help keep it clean but isn't a cure-all, I promise.

If you have a large flock of chickens that would empty your waterer too quickly, consider investing in an automatic waterer. For most backyard farmers, a self-waterer will be good enough. You fill a large water reservoir, which is then turned upside down into a shallow dish that fills with water as the chickens drink down the available water. A word of caution—make sure the vacuum seal is tight on the waterer before you walk away from it or you'll end up with 5 gallons of water flooding the chicken coop. (Yes, that is the voice of experience speaking.)

THORNY MATTERS

In the summer, especially in hot areas, you must check the waterer on a regular basis. During heat spells a chicken will drink three or four times the amount of water as normal. Sometimes in the summer, especially during heat spells, we'll add a water dish in the outdoor yard as well as the water dish in the chicken coop. In the winter months you want to make sure the chickens have access to fresh water that isn't frozen. In our area the chicken coop provides insulation enough to keep the water drinkable. If your area experiences freezing daytime temperatures you can add fresh room-temperature water during the day. Alternatively, you can invest in a heated waterer, but that is certainly not as economical.

Litter management. Chicken manure is very strong, high in nitrogen, and has the potential to smell when it builds up. In fact, one of the primary complaints against chickens comes because of the horrible smell produced by commercial chicken houses. If you do not overcrowd your chickens in filthy conditions, you'll find that the smell isn't noticeable at all.

We use a thick layer of straw and dried leaves in the flooring of the chicken coop. Because our chicken coop is raised off the ground with waterproof flooring, we have to clean out the chicken coop on a regular basis. Once a week we top off the straw with a fresh layer on top and clean out the nest boxes and areas under the roosts where manure builds up faster. About once a month we completely clean out the coop and put the soiled straw into the homestead compost heap.

If the coop were on a dirt floor we wouldn't need to change it out very often, but merely add lots of new hay on top. The bedding at the bottom would, in essence, compost into the soil below. Twice a year you could remove the composted manure and lay down a new deep layer of bedding. In his book *The Small-Scale Poultry Flock,* Harvey Ussery suggests using 12 inches of bedding.

A day in the life. We generally turn our chickens loose after breakfast and let them scrabble about throughout the day. If we need to move the ark, we move it before turning them loose for the day. While loose in their outdoor enclosure, the hens have a grand time. Within a few hours on new ground they have a dust bath area scratched up for their bathing use. They catch any stray bugs that wander in, and dig up any grubs or earthworms that are close enough to the surface to fall prey to their quick eye.

We leave the door to the coop open at all times so they can access the nest box and water containers as necessary. Throughout the day, various hens will visit the nest boxes and the cackling and hollering that emerges from within is a clue that they are working for us. In the afternoons we collect the eggs, check the waterer, and scatter a handful of grains or seeds if needed.

As the sun gets lower the hens make their way into the chicken coop. We don't have a rooster to herd them indoors all together, so the hens tend to wander in one at a time. There's some tussling for the best roost spaces, and my buff insists on roosting on the nest boxes. In the evening after dinner we close the door and windows for the night so they are protected from potential predators.

Signs and Prevention of Illness and Disease

I've heard some say that chickens are sickly and prone to diseases. This has not been our experience; even with the Cornish Cross we haven't had huge loss from disease. When you give your chickens plenty of space and good-quality foods that are fresh and nutritious, you maximize their health potential. Having said that, there are a few things to watch out for in your backyard farm.

Pecking. The dreaded plague you hear about where chickens will literally peck each other to death isn't something that should ever occur on a backyard farm. Debeaking chicks (cutting the tips off their beaks—standard practice in commercial chicken operations) isn't something you'd ever need to do because you will give your flock proper care and feeding. If you have a flock that is turning on each other, you have a flock that is overly stressed, overly crowded, or way too bored. When a flock has enough room, a good dust bath, and a new plot of grass to tear up, they won't be thinking about pecking each other.

Egg eating. Occasionally you will have a situation with a hen that is eating eggs. This usually happens after an egg accidentally breaks (a good reason to maintain proper bedding levels within the coop and nest boxes) and the hen eats the egg. Once a hen develops a taste for it, it's unfortunately all too common for that hen to progress to breaking the eggs open on her own. My father-in-law says only half-jokingly that the best cure for an egg-eating hen is the stew pot. There's usually not a cure for this behavior, so the hen will need to be removed from the flock.

Respiratory problems. There are two main causes of respiratory problems that I've seen in chickens. One is caused by getting damp and chilly, especially in the winter if the chicken coop gets waterlogged or mucky. The second cause is a buildup of ammonia from the chicken manure. If you open the door to the chicken coop and it smells, it is already too strong for your chickens and needs to be cleaned out right away. Provide fresh air and good ventilation, as well as protection from the damp chills. Make sure that manure isn't soggy and too strong by laying down plenty of straw.

THORNY MATTERS

One unusual cause of respiratory problems you might not expect are gnats that fly or crawl into the chickens' sinus cavities at night and suffocate them. While we haven't had this problem in our flock, if you have a chicken die unexpectedly, check to see if you have a gnat problem. Good ventilation will help, and you can spray your window screens, or areas where gnats are seen flying, with a soap and water mixture to kill the gnats when they try to fly in.

Parasites (internal and external). Dust baths are one of the best provisions against a major infestation of mites or lice. Mites and lice will weaken the flock by drinking their blood and making them more prone to illness. With regular dust baths and quarantining newcomers to make sure they aren't infested before introducing them to the flock, you shouldn't have too many problems. Medicated powders to repel mites can be sprinkled in the nest boxes or dust bathing area if absolutely necessary, but we haven't had to use any of the harsh chemicals in more than 10 years.

Internal parasites or worms are minimized by rotating your chickens on their pasture areas so they don't overgraze in any one area. There are deworming medications that can be added into their water, but we use them very rarely in our personal flocks.

Merek's disease. This virus causes weight loss, diarrhea, and often death. There's no cure for the virus, but there is a vaccine that can be given to the chicks. Merek's disease is spread through dander and dust from infected chickens, which means it is primarily seen in flocks that are closely crowded. We don't vaccinate our chicks as we don't overcrowd our flocks and expose them to other flocks.

Eggs from Your Chickens

The boon of the backyard flock is the perfect little gifts offered so humbly for us. The eggs that our chickens produce give us one of the best sources of protein on the farm.

Most eggs produced by dual-purpose breeds have brown-tinted shells. A few breeds seen on the ranch lay white eggs, while only a couple breeds lay blue-tinted eggs. These three shell color variations, white, brown, and blue, allow a wide range of rainbow colors including green and pink tones.

EGG COLOR AND NUTRITION

There is a myth circulating about that brown eggs are more nutritious than white eggs. There's also a rumor that the blue eggs are lower in cholesterol. This simply isn't true. The nutritional content of an egg is not determined by the color of the shell, but by the health and good diet of the hen. Perhaps these rumors started because farm-raised hens were more likely to lay colored eggs. Studies have shown that eggs laid by pasture-raised hens are more nutritious than those laid by conventionally raised hens, regardless of shell color.

Eggs can also appear speckled and spotted. Sometimes eggs are misshapen or have bumpy calcium deposits. Usually these mean nothing; however, if you notice rough bumps on the shells of eggs laid by more than one hen, it can be a sign of dietary imbalance. Add calcium-rich greens such as kale, collards, and broccoli to their diet and give free choice access to crushed oyster shells.

You can expect your hens to begin laying when they are between 20 and 28 weeks old. Pullet eggs—the small "beginner" eggs—are sometimes misshapen, smaller than normal, and might not even have shells. Your laying hens will usually produce the most eggs within the first year of her life and the production decreases gradually each year after that. In dual-purpose and consistent breeds, the decline in production isn't as sharp as what commercial egg layers experience.

After the first year, your hen will go into a period of molting—when they lose their first-year feathers and new ones are formed. Some hens molt with grace, shedding a few feathers here, a few feathers there. Others go completely bald in places, some on the back, on the tail, and so on.

Your hens will stop laying eggs when the days get shorter. If you want to keep them laying more frequently you can provide artificial light in their chicken coop. Otherwise, you can enjoy the natural cycle of their yearly process. In northern climates with shorter days, artificial lighting might help extend the laying season. Focusing on breeds that are winter-hardy and tend to lay better on a year-round basis can help backyard farmers in these regions keep a steadier supply of eggs.

As your flock ages, your egg production will naturally taper off. Bring in new baby chicks each year, or every other year at the most, to keep your laying flock healthy. In the backyard flock with broody hens, this can be as simple as letting one of your hens "go setting" and hatch out a few chicks each year.

We separate the brood hens from the rest of the flock. This pen may seem small, but a broody hen rarely leaves the nest and wants a close, private-feeling space. The tarp can cover the front during a storm, and the pitched roof sheds light rain without trouble.

For details about building a simple brood box, see Appendix A.

There are as many ways to tell if a hen is laying well as there are farmers. One of the most consistent ways I've used to really tell, is to check the space of the pubic bones. Pick up your hen and find the two pubic bones just above the vent. Press your fingers into the soft space between the two pubic bones—if the bones are two or three finger widths apart, that's a good sign of a consistent layer, as nonlaying hens will generally have close pubic bones.

With a small mixed-breed flock, it is very likely you will learn which hens are laying simply by the color of the shells. Each egg may have a distinction that makes it easy to tell them apart. Obviously, if your eggs are more uniform or your flock is large that isn't possible. One backyard farmer we know chooses a different breed of chicken each year so he can tell the production levels according to the color of the eggs. Generally speaking, by the third year of life even the most consistent breeds won't produce enough to justify the cost of feed.

Collect your eggs once a day at a minimum, twice a day if possible. Brush off any loose straw and store the eggs in the refrigerator right away. They will store longer if you don't wash them. If you wash the eggs you will remove the bloom, a protective membrane covering that helps prevent bacteria from getting through the porous egg shell. I only wash eggs that are soiled with droppings and make sure to use those eggs first because they've lost a level of protection.

Eggs are the primary reason for keeping chickens in the backyard farm. With more and more cities lifting restrictions against backyard flocks, there's no reason not to enjoy fresh eggs right from your own sweet hens.

ROOSTERS?

There is a myth that your hens lay better eggs if there is a rooster available. This isn't true. Your hens will lay eggs whether there is a rooster in the flock or not. However, in the backyard flock a rooster will allow you to hatch your baby chicks each year because the eggs will be fertile. A watchful rooster will also help protect your hens, keep them together in one area, and help find forage for them.

A good rooster will watch over your flock, keep the hens out of trouble, and fertilize your eggs so you can hatch your own chicks. It is common for a rooster who's uncovered a hidden trove of food to immediately call over his harem and stand proudly watching as the hens take first dibs.
(Photo courtesy of SammyDavisdog)

Meat from Your Chickens

It was quite a mental journey for me to go from harvesting eggs from my chickens to harvesting meat from the chickens. It helped that my husband knew what he was doing. It also helped to do some study on the commercial processing of most industrialized chicken houses.

Studies show the meat of pasture-raised poultry is significantly more nutritious. For many backyard farmers it is less expensive than store-bought meat as well, especially as the prices rise. Never mind the cruelty-free factor, which is easier to confirm when you raise your own meat. Like my husband says, "The best way you can know for sure where your meat comes from is to raise it yourself."

The biggest factor for me was to not name the chicks that were destined to be butchered. In fact, I called them a collective "The Chicks Who Shall Not be Named," but then I'm weird like that. I think mentally preparing yourself for the idea that *these* are chickens destined for the freezer is helpful. I'll be the first to admit I was a little squeamish the first year, but the feeling of pride that comes with the knowledge that you've provided a year's worth of the highest-quality poultry for your family is amazing. It spoke to my self-sufficient and independent streak.

When we raise Cornish Cross for butchering we do not feed only high-protein, artificial feed mixtures, but rather allow our Cornish Cross chicks to free-range in pastured areas. Instead of reaching full size in the 7 to 8 weeks as commercial chicken houses experience, we butcher our chickens in about 11 to 12 weeks but have only used half the "recommended" feed amounts. This keeps the cost of feed down, and makes sure that our chickens are able to enjoy a healthier lifestyle for the few weeks they are with us.

We still dressed out an average of 5-pound carcasses to put in the freezer, so we have had good success with this method. Last year we purchased 40 *straight-run* chicks (a mix of *pullets* and *cockerels*), received 42, and raised 38 to butchering size. The slightly slower growth rates seem to keep our chicks healthier and our death rates are very low.

DEFINITION

When you order baby chicks as **straight-run,** you are getting a mix of male and female chicks because they haven't been sexed, or sorted by gender. A **pullet** is a young female chicken; a **cockerel** is a young male chicken.

We fed about one 50-pound bag of feed a week at $7 average per bag of feed. The cost of feed was about $84 (this price includes feed used for the laying pullets as well). The cost of our chicks was about $50 with shipping and handling. The total cost spent was less than $150 total for the 38 birds we put in the freezer. The chickens were an average of 5½ pounds dressed out, giving us roughly 200 pounds of meat. With grocery store prices at more than $2 per pound for industrially raised chicken, we saved more than 75 percent on the price of chicken and enjoyed a higher-quality meat that was raised in more humane conditions.

When I truly understood the filthy and sickening conditions that the average meat bird is raised in I was able to set aside my momentary discomfort a couple times a year to do my own part to avoid contributing to this system. I never take it for granted, and certainly don't find pleasure in killing animals of any kind, but rather I am grateful for the ability to provide for my family and for the provision we enjoy from our flock. I hope that makes sense.

I will discuss butchering techniques in Chapter 16. If you choose not to raise meat birds for yourself, I hope you'll make purchases from locally raised poultry and support local backyard farmers.

Other Types of Poultry

Depending on the size of your backyard farm you might enjoy trying your hand at some of the other types of poultry. Geese, ducks, guinea fowl, and turkeys have all found an excellent place in the homestead.

In all of these poultry types, heritage breeds that were once common on farms and homesteads are still available—and in many cases, better suited for a backyard farm than the most common commercially raised strains. The American Livestock Breeds Conservancy is an organization dedicated to preserving information about these delightful heritage breeds. You can find out more about specific breeds at their website: albc-usa.org.

Guinea Fowl

These large birds are considered excellent watch dogs around the homestead. My husband jokes that you cannot sneak up on a house with a flock of guinea around! They are rather noisy and can fly well so they will often roost in trees instead of sheltering in chicken coops when given the chance. For this reason, if you have neighbors close by you may not be able to keep both guinea and neighbors. Guinea are fabulous for pest control, however, and will eat slugs, ticks, and a host of other ickies. They lay large eggs on a moderately regular basis and are excellent foragers.

Geese

Equally large and noisy, geese will warn you about snakes, intruding cats, and all kinds of other potential threats both real and imagined. Geese do lay very large eggs but there aren't any geese that match even ducks, nevermind chickens, in the number of eggs produced. As with guinea, geese are great foragers and will attack weeds with delight. You can use geese to help clear areas for planting before tilling them up and sowing a new crop. A goose was at one time a traditional Christmas dish. They are also a source of down feathers for warm comforters and pillows.

Ducks

Ducks, like geese, are excellent foragers and will eat slugs and grubs with relish. Some breeds are known to be excellent egg layers, while others are grown more for their meat. While ducks have a reputation for being noisy, one of the most popular homestead breeds, Muscovies, are near-silent and better suited for backyard flocks. If you live in an area with a pond, ducks will keep mosquito larvae at bay. Ducks do not make reliable mothers compared to broody hens, so many homesteaders who want to hatch out their own ducklings will use a broody hen for hatching ducklings.

Turkeys

Turkeys are considered too hard to maintain and picky to raise for many homesteaders. However, heritage breeds are hardier and generally make better foragers. A backyard farmer who wants to be self-sufficient should make sure they start with a breed that can breed naturally and has some brooding instinct still intact.

THORNY MATTERS

Did you know that 99 percent of all the breeding turkey stock is held by three main companies? Did you also know that these commercial strains are almost entirely incapable of breeding on their own?

One of the biggest problems you'll face starting out with new baby poults (young turkeys) will be "starving out." In some strains the natural foraging and feeding skills are so lost, the poults will stand beside a full container of feed and starve to death. One way to handle this is to put week-old chicks in with your new day-old poults so the older chicks can literally teach the turkeys how to eat and get them started. Remove the chicks before their turkey roommates grow too large and start bullying them around.

Rabbits on a Backyard Farm   ∞ 13

Even in the smallest of spaces, rabbits can be kept in a backyard area with ease. Some people raise rabbits for meat, others for the fiber that they can spin into yarn, and still others simply keep them as pets or for their compost-enriching manure. Whatever your reason for keeping rabbits, most areas consider rabbits as pets for zoning purposes, giving backyard farmers in restrictive cities an option for livestock raising where chickens or goats might be banned.

They are quiet and, if their manure is managed properly, odorless. In fact, rabbit manure comes in the form of easy-to-handle pellets. Bunny berries, some people call them.

Rabbits for Meat

There are many types of meat rabbits and they have been bred for many years for a reliable source of meat production. Their meat really tastes good and is high in lean proteins. Rabbits are extremely reliable and in some areas they are a meat staple.

Rabbits will dress out at about 6 pounds of meat. They also convert feed more efficiently than cattle will, with rabbits producing 1 pound of meat for every 4 pounds of feed, while cattle takes 6 pounds of feed on average per 1 pound of meat.

Three or four *does* in prime health produce enough meat to support an average-sized family's protein needs for a year. A doe needs to mature about 7 months before being old enough to breed, and 6 to 7 months old for a *buck*. Each breed will have slightly varied time of maturity.

DEFINITION

A **doe** is a female rabbit; a **buck** is a male rabbit. A castrated male is called a *lapin*. Baby rabbits are called *kittens* (often shortened to *kits*), and a young female that is not yet mature is sometimes called a *doeling*.

Altex. Altex rabbits in maturity weigh from 10 to 20 pounds and originated from Texas. They are purposefully bred for rabbit meat production and are often used as sires to breed with does of other breeds. They are rapid and efficient in weight gain, with a high dressing percentage, and a high meat-to-bone ratio. The does do not tend to make great mothers so are most often used in partnership with other breeds.

American Chinchilla. The large, meaty type of the chinchilla furred rabbits, the American chinchillas mature between 9 to 12 pounds. These rabbits tend to have large, successful litters and are good tempered to handle. Modern meat strains have a quicker growth rate to reach a size suitable for butchering.

American Rabbit. Bred both for meat and fur, this breed of rabbit originated in Pasadena, California. The mature weight is between 9 to 12 pounds and the breed is considered endangered. Enthusiasts love the breed for its gorgeous blue fur and good temperament.

Beveren. This breed originated in western Belgium. Litters are large, and the young grow fairly fast, making them a good converter of feed to meat. Beveren does are typically docile and make good mothers, and in some areas they are used as a large pet breed because of their gentle disposition. The Beveren is a hardy breed that has a mature weight of 8 to 12 pounds.

Blanc de Hotot. The mature weight of the buck is between 8 to 12 pounds, and the doe weighs from 9 to 11 pounds full-grown. An active, hardy breed, the Blanc de Hotot is highly attractive with a glossy white coat and thin black coloring around the eyes. They are fairly good mothers, have good-sized litters, and the young grow rather fast, making them an excellent choice for meat production. Blanc de Hotots are considered an endangered breed.

California White. This is one of the primary breeds of meat rabbits in the United States today and is popular for the quick growth of the baby rabbits. The bucks mature between 8 to 10 pounds, while does generally reach mature weights of 9 to 11 pounds. Many breeders are now experimenting with crosses to help lend the California White's faster growth rates with the size seen in some of the other breeds.

Checkered Giant. This rabbit breed has well-defined, black and white markings. The mature weight of the buck is 11 pounds, and 12 pounds for the doe. These are not considered one of the most common meat rabbit breeds because purebreds can take longer to mature, and are prone to aggressive behavior. Some meat breeders will use one-quarter crosses as breeding stock to try to increase mature size.

Cinnamon. Cinnamon rabbits are a rare-to-find breed that originated in Montana. Their mature weight is from 8 to 11 pounds. A hardy breed, it usually lives between 5 and 8 years, and has a calm, easy-to-handle demeanor that enjoys attention. They have a commercial meat-type build and are good for meat production.

Flemish Giant. This large-sized rabbit, which weighs from 12 to 14 pounds in maturity, came from the Flanders region. It takes 1 to 1½ years to reach their full maturity. Pure-breed Flemish Giants aren't typically used in meat production because of the long maturation times, but crosses are used with care.

New Zealand. This breed, along with Californias, are now the top meat rabbit breed. Bucks weigh 9 to 11 pounds at maturity, and does weigh 10 to 12 pounds. The large litter sizes and meat-purpose focus of this breed have led to selective breeding for traits you want in a meat rabbit: large litters, fast growing kits, good mothering skills, and good feed-to-meat conversion.

This white rabbit is a good meat type. You can see why New Zealands are popular as a meat rabbit, or as a pet for younger children because they are sturdier for little hands to hold.
(Photo courtesy of Chris McLaughlin)

Palomino. Palominos are less common meat rabbits, but the bucks weigh 8 to 10 pounds and does weigh 9 to 12 pounds. Originally from Mexico, they are also raised for their fur. Palominos have smaller bones for their meat ratio, so they are still used in meat production even with a slightly slower maturity rate.

Silver Fox. The Silver Fox bucks mature at around 9 to 11 pounds, while the does average 10 to 12 pounds. Does have large litters, produce plenty of milk, are excellent mothers, and even make wonderful foster mothers. It was the first of the larger breeds to dress out at 65 percent of live weight. Considered a threatened breed in the United States.

Rabbits for Fiber

Angora rabbits are relatively simple to care for. Keeping all of them properly groomed may be the greatest chore, but for fiber lovers who want a source of fiber for spinning, the time is well worth investing. The benefit of a breed like Angora is that you can brush out the fur as it sheds without having to kill the rabbit. The angora wool can be hand-plucked or sheared (see "Harvesting Fiber from Rabbits" later in this chapter).

Long fibers that have been hand-plucked contain the full fiber length end-to-end and are often sold at a premium price to spinners because they are easier to spin. Shorter fibers or fibers from the breeds that need to be sheared are easier to harvest via shearing, but bring a cheaper price sold to spinners. Angora wool produces a very warm and insulating yarn, and is sometimes mixed with wool from sheep or alpaca to make it easier to spin and knit with. On average, one shearing produces enough fiber for a scarf.

OVER THE GARDEN FENCE

Rabbits raised for fiber need regular grooming and maintenance. A thorough brushing at least twice a week will help prevent mats, tangles, and wool block. Clip nails once or twice a month, as needed.

English Angora. Considered one of the cutest types of Angora Rabbits, the English Angora has long, thick fur that almost covers their face and ears. They have very silky, soft, and fine white wool, which is sometimes considered one of the best colors. The English Angora also comes in a range of other magnificent colors, but colored wool is more difficult to dye. This breed of rabbit has a mature weight between 5 to 7½ pounds. The longer facial fibers means that you'll need to stay on top of regular maintenance routines.

French Angora. Weighing 8 to 10 pounds at maturity, French Angora rabbits have no wool on their face, ears, head, and feet. This is known as a clean face because the fur is a normal length. This is the easiest type of rabbit to care for due to its wool's higher proportion of guard hair to under wool. The fiber qualities are excellent, making it easy to spin and giving the wool a nice fuzzy bloom.

German Angora. With the large amount of wool fibers this type of breed can produce, it makes them popular among handspinners. German Angoras aren't a recognized breed in the United States, but they are popular for fiber growth. German Angora has a mature weight between 9 and 12 pounds.

Giant Angora. The largest of the Angora breeds is the Giant Angora, which has been used commercially because of its large size and excellent wool production. They look similar to the German Angora but have a mature weight of nearly 20 pounds. This breed has three fiber texture kinds and the most dominant is the under wool, which is very dense and must be sheared.

Satin Angora. The mature weight of the Satin Angora is 8 pounds. Satin Angora types have shiny wool fibers but don't have wool on their head, face, ears, and front feet. Their wool feels much lighter compared with other breeds and has a high-gloss sheen. The grooming routines have to be maintained on a regular basis.

This Satin Angora has a clean face, but lots of long wool for fiber production.
(Photo courtesy of Steven Depolo)

Rabbit Husbandry

Taking care of your rabbits is fairly simple, but as with any pet, the basic needs should be met in a timely fashion. Rabbits are clean, when given proper housing arrangements, and are usually quiet. In an area with nosy neighbors, you might find that rabbits are the animal least likely to cause a fuss.

Housing Rabbits on a Small Scale

Each rabbit will need a cage or hutch of its own. I know you often see rabbits housed together in a group in one large cage, but this can lead to problems with indiscriminate breeding and even inbreeding. You want to control the breeding that takes place in your rabbits and have a place to remove the males to when the mother gives birth. As with chickens, rabbits can be easy prey for predators, so make sure your cages are built out of sturdy materials, and with protection from inclement weather to boot.

While I have seen backyard farmers who keep their rabbits completely free-range, I also hear the horror stories about the massacres that happen when a hawk or stray dog finds the buffet. Even in urban areas, there are animals that will prey on your rabbits if given half a chance. There's also the threat that your rabbits will dig under a fence, get loose, and be eaten, lost, or even struck by a car. Provide safe housing for your rabbits and you'll be in good shape.

Rabbits need a space of at least .75 feet per pound of expected mature body weight in their cages. Make the top of the cages at least 18 inches tall, more for larger breeds of rabbits. You'll want to make sure that the cage is comfortable for the rabbit and doesn't have any sharp edges to cut their thin skins.

THORNY MATTERS

Wire is the typical building material for rabbit cages, but any sharp protrusions from where the wire has been cut can harm your bunnies. Check inside the cage, including door openings, with your hand and file down any sharp edges.

Cages should provide good ventilation for your rabbits so that fumes from their manure don't build up and harm their eyes or lungs. Wire is often used, which gives the added benefit of being self-cleaning, allowing the droppings to fall out of the cage and onto the ground or floor below.

Rabbits tend to prefer cooler temperatures, around 55°F to 60°F, so climates that get warmer will need plenty of shade and protection from direct sunlight. It's easier to raise rabbit stock that has been bred in these warmer climates than to ship rabbits from milder climates to hot southern zones where they will be stressed by the transition. Even the hardiest breeds will need protection from wet weather, chilly drafts, or overheating.

Any cage or hutch that you put together should also be easily accessed for daily maintenance. You'll need to clean out cages, provide fresh water and food on a daily basis, and check on the welfare of your rabbits. If the cages are difficult to get to you'll be creating extra work for yourself, so save the hassle and make or purchase easy-to-use cages right from the start.

The wire you use in your rabbit cages or hutches should be a sturdy-gauge wire that won't be easily broken through. The squares should be large enough for good ventilation and to allow pellet droppings through pretty easily. But the wire squares should be small enough to prevent kits from slipping through the openings. Some breeders use a type of wire called "baby saver" wire, which has smaller openings at the bottom of the cage sides to keep the babies from crawling out.

Chicken wire in the typical woven hexagonal pattern is not strong enough for a rabbit cage. Try welded mesh wire in a 14-gauge strength and make sure the size of the squares is around $\frac{1}{2} \times 1$ inch to allow plenty of support for the feet of your rabbit. Any wood portions of your cage or hutch should be inaccessible to your rabbits because chewing treated lumber can make them sick.

Raised rabbit hutches will hold more than one cage. These structures will put the cages at a level where caring for the rabbit is easier. They also allow droppings to fall down to the ground below which many backyard farmers will use as a place for raising worms. These worm beds under the rabbit hutches can provide rich worm castings and great soil amendments for the garden while eliminating any potential odors from the rabbit droppings. A true win-win situation—which is what I love to see on the backyard farm.

This rabbit hutch allows for droppings to fall right into the worm boxes below. This provides good sanitation for the rabbit and helps make fabulous vermicompost at the same time.
(Photo courtesy of Chris McLaughlin)

Care and Maintenance

Rabbits need good-quality feed, especially when you're raising them as livestock. Both meat rabbits and Angora rabbits need to have a good-quality feed to put on the weight for meat or to produce high-quality wool. Lots of fiber and good protein will be needed. Rabbits can eat lots of the veggie trimmings from your garden, plant thinnings, or kitchen scraps.

Many breeders keep a high-quality hay available, such as timothy grass or alfalfa hay, for fiber and protein. Chewing on the hay helps keep their teeth trimmed down, as rabbit teeth will grow continuously and need to be kept in check. In addition to the hay, they should be given enough feed each day to meet their other dietary needs—anywhere from 4 to 8 ounces for young rabbits, and up to 8 to 11 ounces of feed for pregnant does.

Some breeders will feed straight rabbit pellets that are 18 percent protein, while others will feed only scraps and other vegetable matter from around the backyard farm. Most likely you'll find yourself somewhere in the middle, feeding a portion of whatever fresh greens you have available and filling in the gaps with pellets. Remove any of the fresh foods that haven't been eaten at the end of the day. This prevents bugs and mold from growing.

Your rabbits should never be allowed to get too thin. Allowing your rabbits to get too obese can cause a whole different set of problems. Handling your rabbit on a regular basis will help you judge if they are getting enough feed. It can be tough to tell, especially with the Angora rabbits, just by looking at them, so be sure to feel their body weight and see how they are doing.

Rabbits need constant access to clean fresh water, so you'll want to check their water dishes every day. Clean out the water dishes every couple of days, or whenever they are soiled by food or droppings. If your doe is nursing a large litter, be sure to check her water more often—allowing her water to run out is a big no-no.

Regular cleaning and sanitizing of the cages is needed as well. This helps prevent buildup of fur, debris, and manure, which can attract bugs, bacteria, or diseases. Clean out the obvious debris with a wire brush, and then sanitize the cage with disinfectant. Full sanitization should take place between every breeding cycle or regularly every month or so.

Breeding your rabbits should be done only when you need the offspring. Place the doe in with the buck, watching carefully in case of fighting. Allow them to breed for 15 to 20 minutes and then remove the doe. Rebreed the pair after an hour or two as the second session increases the litter rate. Then remove the doe back into her cage. Does progress quickly and usually have their babies on the thirty-first day. Add a nest box to the cage at the twenty-fifth day so she can begin to build her nest. When you notice the doe pulling fur to line the nest box, you should have kits within the day.

Avoid disturbing the nest box for the first couple days after the kits are born except to remove any deceased babies. Watch out for any kits that have slipped out of the nest box, though, as sometimes when the doe leaves the nest they will ride out latched on to the nipples. Does will not retrieve their young, so you'll need to place them back in the nest box if they are still alive. Record any litter sizes, both live and dead kits, so you can track your breeding efforts.

Many breeders use tattoo identifications for their livestock. Simple home kits are available so you can tattoo your rabbits on their left ear. This allows you to better track their daily care, breeding, eating, and so on. This is especially important for rabbits you plan to keep as breeders, or sell to others for use as breeders.

By 14 days old, your babies should have their eyes open. If they don't, you may need to use a Q-tip and warm water to clean their eyes gently on a daily basis until their eyes open. Most breeders remove their kits from their mothers around 6 to 8 weeks of age.

CHEWS AND ENTERTAINMENT

Provide some toys or entertainment for your rabbits to avoid behaviors like aggressiveness or fur chewing/plucking. You might create an outdoor ark or bunny-run to allow them time outdoors in a protected area. Other times you can add chew toys like the wood chews for large rodents from the pet store. A homemade source of chewable wood that is safe for rabbits are the tips of apple trees pruned from your home orchard. However, never give your rabbits apple branches that you don't know the source of. Nonorganic methods of growing fruit involve a huge amount of chemical sprays and applications that can make your rabbit sick. Only provide apple branches from sources you are familiar with and can trust.

Signs and Prevention of Illness and Disease

Many diseases can be avoided by good-quality food, daily care and maintenance, and proper housing for your rabbits. However, just as with any animal, sometimes your rabbits can get sick. That's why daily checks on your rabbits' demeanor, not just filling food and water dishes, are important.

Handling your rabbits often will make it easier to tell when your rabbit is sick. This girl enjoys her Angora rabbit as a fun 4-H project, as well as a beloved pet.
(Photo courtesy of Steven Depolo)

Ear mites. Although they are especially common in overcrowded or neglected rabbits, any rabbits can contract these annoying pests. When you see signs of ear mites, you need to take action right away to prevent disease spreading and complete infestation. Mineral oil can sometimes smother out the mites if the infestation isn't too far advanced. Otherwise, a prescription ointment may be needed from your veterinarian. You'll need to thoroughly clean any cage with a mite infestation, as well as the entire hutch area. Mites are notorious for hiding the crevices of a wood hutch and being hard to completely eradicate.

Enterotoxemia. This potentially fatal problem that can occur with rabbits happens when their digestive balance is disturbed. This can happen when too much fresh fruit or greens are given at one time. Sometimes you may notice signs of diarrhea in your rabbit and be able to give them more hay and uncooked, whole oats, while other times you may lose your rabbit before you ever saw anything wrong. Any changes in feed should be made slowly, especially during already stressful times such as weaning, breeding, or moving to a new home.

Giving your rabbit some probiotics such as Benebac gel can help improve digestion when you don't otherwise know the cause of the diarrhea. Many breeders will give probiotics anytime antibiotics have been administered as well, because antibiotics will destroy the healthy gut flora that helps with digestion. Be sure to provide plenty of water, remove feed rations, and give the highest-fiber foods like hay and whole oats to give your rabbit the best chances at recovery.

Heat stress. This can be a serious problem for rabbits in warmer climates. Keep good ventilation in your housing to help them maintain their body temperature, and never let their water run out. Make sure their houses are turned out of the direct sunshine on hot summer days. Some breeds, like the Angoras, need proper grooming to avoid mats that will increase their body heat during the hottest days of the summer.

OVER THE GARDEN FENCE

In the warmest climates, you might try an old trick that works great—freeze water in leftover water bottles or pop bottles and then place a frozen water bottle right into the rabbit cage. This allows your rabbit to have a cool spot to snuggle up on during the hottest hours of the day.

Parasites. Anytime you see loose stools you should check for the potential of parasites. Internal parasites can sometimes spread from bunny to bunny. If you have a larger rabbitry with several tenants, you want to identify the parasite and treat quickly to avoid spreading the disease. Obviously, good sanitation will be a must in these cases as every cage and piece of equipment will need to be sanitized, as well as hand-washing routines put in place in between the handling of each rabbit.

Pasteurella. This disease causes coldlike respiratory symptoms in rabbits. If you notice your rabbit sneezing a lot, any white mucous draining from the nose or on his paws from where he's been wiping his nose, or weeping eyes from blocked tear ducts, chances are he has a *Pasteurella multocida* infection, or pasteurellosis. Pasteurella is a bacteria that almost all rabbits are exposed to; however, only those with weakened immune systems will usually develop symptoms. Stress from weaning, pregnancy, moving, overcrowding, poor living conditions, and other preventable situations can trigger pasteurellosis.

Sometimes antibiotics will help prevent secondary infections, but there is no true cure for the disease. That's why good breeding habits and good living conditions are so important. You should never breed a rabbit that has shown symptoms of pasteurellosis, although offspring from a sick doe that stay healthy to a breeding age may have a higher-than-average immunity.

Harvesting Fiber from Rabbits

Angora rabbits are a great source of fiber for the backyard farmer who is also a spinning enthusiast, or for someone who is simply looking for another avenue of income from their homesteading endeavors. Not that you'll get rich by selling your rabbit fiber, but you might make enough back to help cover your feed costs.

German Angoras have been bred for many generations to hold on to their hair so they do not shed. This means you have to shear their wool in order to harvest it. The other breeds of Angoras will usually go through regular molts, at which time you can pluck the fibers by hand. This is more time intensive than shearing, of course, but is thought to produce a longer, easier-to-spin fiber.

To hand-pluck your rabbit, wait until you notice the rabbit beginning to molt. You should notice the new fur growing in underneath the longer, older hair that is beginning to shed. Then follow these steps:

1. Get comfy in a chair (some prefer to place their bunnies on a table top or counter where they can stand comfortably) with a towel or blanket on your lap. This will protect you from any accidental scratches and give your rabbit a more firm-feeling surface to stand on.

2. Grasp a small section of the longest hairs, about ¼ to ½ inch wide so that you aren't pulling on too much at a time.

3. Gently pull on the hair until it releases and then grasp a new section.

4. Continue to gently pull out the long hair until only the new incoming hair remains. Keep three buckets or boxes near you while you do this so you can put the longest, prime-quality fibers in one place, the shorter "seconds" in another, and the matted, dirty, or unusable short fibers in the third throw-away bucket. This will save you the trouble of sorting the fiber later.

To shear your rabbit, you can place him in your lap or on the table top as before. It's often useful to use the cool setting of the hair dryer to blow out any hay or debris. Brush the hair smooth to make sure there are no mats. Then you're ready to cut the wool.

Some people use electric shears similar to sheep shears. Others use scissors like small sewing scissors that are easy to hold in your hand. If you use electric shears you'll save a lot of time, of course, but that doesn't mean to use less care. Special care should be taken to avoid clipping the skin so you don't hurt your rabbits.

Look at the pile of high-quality wool that can be harvested from a single shearing.
(Photo courtesy of Chris McLaughlin)

THORNY MATTERS

When clipping does, special care must be taken when clipping their belly. Make sure you find the teats and use scissors to clip a circle around them so you won't accidentally hurt them with the shears!

When shearing the dewlap (the loose skin under the chin) and neck area, use caution. Lots of folds of skin can make it difficult to trim this area. Don't be afraid to do the shearing in more than one session—especially if this is the first time for you or your rabbit. Let them stretch, relieve themselves if necessary, or get a special treat for being so good during your time together.

The biggest key to successfully harvesting the fiber from your Angora rabbits is a bond of trust between you and your rabbit. Handling your rabbit early and often is the best way to make that happen. When your rabbit is comfortable being handled by you, your harvesting will be much easier, regardless of which methods you use.

Sheep and Goats: Backyard Multitaskers

14

When I began looking for ways to decrease the amount of milk we purchased through commercial dairy systems, I started researching dairy goats. What I discovered is that goats, and their cousins, sheep, can be huge workhorses in the backyard farm.

Smaller than other hoofed livestock like horses or cows, they provide many benefits. If you have room for a couple goats, you can greatly increase your self-sufficiency. Goats and sheep can provide milk, fiber, and meat for even city homesteaders. We live within city limits on a suburban corner lot and have been keeping goats successfully for years.

Breeds for Small Home Use

While commercial operations will usually keep breeds that are specifically raised for meat, or specifically raised for fiber, a backyard farmer might want to keep heritage breeds that are more dual-purpose. This will allow you to gain multiple benefits from a single animal. Here are some breeds you might want to consider for your backyard farm.

Breeds for Milk

Alpine. Alpines have been bred primarily for milk production, but *bucks* should always be more than 170 pounds. This good size means that they are a suitable breed for meat production, which makes dealing with *wether* culls (getting rid of the extra, castrated males) much easier. One of the benefits of the alpine breeds is that they are highly tolerant of poor ground and are hardy—easy to maintain, usually keeping very good health, and great mothers. An alpine *doe* can be expected to produce around 3 liters of milk daily for 9 to 10 months a year.

East Fresian Sheep. These sheep are one of the most popular milk breeds and will produce very rich milk. They are highly productive, with most sheep averaging between ½ and 1 gallon per day for about 200 days a year. And with a mature weight of 150 pounds (*rams* can reach up to 200 pounds!), these sheep can make a good meat choice as well. Purebred East Fresians don't tend to do as well in large, commercial conditions, or in areas with extreme heat, but can be crossed with other breeds to increase milk production. A mature *ewe* will also produce between 8 and 11 pounds of wool per year.

These goats fit nicely into our backyard and not only clear the fence line, but provide meat and milk for our family.

Lacaune Sheep. These sheep have been very carefully bred into highly productive milking sheep that will now produce 200 to 500 pounds of milk each year (from U.S. bloodlines). While the European lines still produce better than those breeding lines available in the United States, there is a great market for sheep cheeses in the United States. They are more heat tolerant than the East Fresians, and their milk has a slightly higher butterfat content. They do not produce any real wool to speak of, and also produce fewer *lambs* compared to the East Fresians.

DEFINITION

A **buck** is a male goat; a **doe** is a female goat. A **wether** is a castrated goat. A **ram** is a male sheep; a **ewe** is a female sheep. Baby sheep are called **lambs.**

La Mancha. We started with La Mancha goats in our herd because they were available in our area and produce a lot of milk. We get between ½ and 1 gallon of milk per day per doe, which provides plenty of milk for my family. La Manchas have a slighter build, though, and usually we breed them with a Boer or Nubian buck so the offspring will be better suited for meat purposes. One of the biggest stand-out features of the La Manchas is that they have very small or even nonexistent ears. Ours are friendly, vocal, and personable.

Nigerian Dwarf. Nigerian goats have a great disposition, and their smaller size makes them popular for a backyard space. Not as well suited as a meat goat because of the small size and thin build, the Nigerian dwarfs are rock stars as milkers. A doe can produce 2 quarts of milk per day and their milk has a higher butterfat and higher protein level than many of the other milking breeds. Because they are so easy to handle, they are popular for 4-H projects, children's pets, or first-time owners.

Our La Mancha milk goat, Ginger, has a more streamlined appearance with an already visible udder though she's still 4 weeks from giving birth.

Nubian. Nubian goats are one of the most popular dairy breeds in the United States right now and can also serve as meat goats. Full-sized goats range between 125 and 175 pounds, so they are one of the larger milk breeds. Nubians have high butterfat and protein content in their milk, which makes them more popular for making cheeses. A Nubian doe can produce 1,400 to 1,500 pounds of milk per year. Nubians are more vocal, like La Manchas, so if you have neighbors close by, that is something to keep in mind.

Breeds for Meat

Boer. One of the primary breeds of meat goats, Boers are large-framed and much thicker than many of the other breeds. You look at a Boer and you can tell it is primarily a meat goat. They are bred specifically for good weight gain, high conversion of feed to muscle ratio, and of course, size and meatiness. Our Boer doe has a big, broad, Roman nose and floppy ears. Boers are actually decent milkers as well and have rich, high-butterfat milk that they use to feed their twins, triplets, or even quadruplets. While many strains have been selectively bred for meat production only, it is possible to find strains of Boers with more productive milking abilities also. One cool thing about Boers is that they are the only goats that can have four functional teats.

Dorper Sheep. One of the fastest-growing meat breeds of sheep, the Dorper originated in South Africa, so it is well suited to hot, dry, and less than ideal conditions. The sheep grow quickly and give a backyard farmer a good return on feed investments with their ability to thrive in a variety of conditions. For someone looking for a very low-maintenance breed, the Dorper or White Dorper sheep do not require shearing. They are sometimes used as crosses for backyard farmers wanting to improve quick growth, good mothering, and a high rate of lambing.

The body size and makeup is obvious when you compare the meat goat breeds with the milk breeds. Our Boer goat, S'mores, has a larger frame, thicker bone structure, and far more meat on her than a typical milk breed.

Kiko. Kiko goats originated in New Zealand and are similar to Boer goats in their size and meatiness. Kikos were feral at their origin so they are very hardy and disease resistant. And with only the most maturing and fastest-growing goats used for establishing the breed, the Kikos rival the Boer goats in healthy feed-conversion rates. Most Kikos are white and have short fur.

Myotonic Goats (Tennessee Fainting Goats). This breed of goat originated in the United States but their common name is slightly misleading. While Myotonic goats do have a trait that causes their legs to stiffen up and sometimes fall over, they remain conscious when this happens. This stiffening action is partially what creates the thick meatiness of the breed, making them a suitable meat breed. Some Myotonics are naturally polled, or hornless.

DEHORNING AND DEBUDDING

If your goats are not naturally polled, they will produce horns as they grow. You can have your goat dehorned or disbud the kids when they are a couple days old. Disbudding involves burning off the growth ring of the horn bud and should be done with help from an experienced goat keeper. Some people choose not to disbud or dehorn their goats. We dislike the disbudding procedure and leave our goats in horns for a variety of reasons. Horns are helpful for holding a goat during vaccinations and other procedures, help with temperature control of the goat, and are more natural for the goat. Commercial dairies all dehorn their goats to prevent injuries from multiple goats in the same enclosures. The choice is a personal one for each goat owner.

Savanna. Savanna goats are a little hardier than Boers, as they originated in South Africa and have more tolerance to dry, hot weather. They are said to be more resistant to parasites and diseases as well. Savanna goats also can develop a cashmere coat that increases the value of the breed for the backyard farmer. If you live in a drought-prone area, or want a dual-purpose goat, you might consider the Savanna.

Texel Sheep. Texel sheep are a popular meat breed in the United States and are known for their quick growth and lean meat. The sheep must be white according to breed descriptions so even though they only produce 3 to 4 pounds of medium-length wool, the wool is useable. They are considered excellent mothers as well and have a high rate of multiple births with twins considered the norm. Texel sheep are very efficient at turning feed into muscle mass and are naturally lean with little fat.

Breeds for Fiber

American Cormo. These fine-wooled sheep originated in Tasmania but came to the United States in the 1970s. Since then they have become known as producers of a fine fleece with more than 6 to 9 pounds of wool produced each year. They are low-maintenance sheep and can tolerate wet or humid climates better than some of the other sheep breeds. They also produce twins often and are excellent mothers, which helps make them easy to keep.

Angora Goats. Angora goats produce mohair, a distinctive long coat with wavy or curly fibers. Originally from Turkey, they are popular with homesteaders for their spinnable fiber. Their medium size allows them to fit in nicely with a backyard farm program. Traditionally, the Angora goats were white, but a growing trend among the breed is producing a variety of colors that allow for naturally colored fibers. Red, grey, and brown are all colors making a comeback among the Angora enthusiasts.

Border Leicester Sheep. Popular for the long, white wool that can have a staple length up to 10 inches, the Border Leicester is often used to cross with other breeds and improve fleece length. The average ewe will produce 8 to 12 pounds of wool each year. While they are not as highly productive in lambing, they are excellent mothers and milkers. Their parent breed, the Leicester Longwools, produced a number of Leicester cousins that all have a similarly long staple length to their fleeces.

California Red Sheep. A newer breed that is still developing, I anticipate seeing more wool production from this breed in the future. Right now, these naturally colored sheep produce 4 to 7 pounds of wool per year. California Red sheep do not have wool around the belly and udder area, and the rams have a red mane of hair around the neck. Their color begins as a dark cinnamon red when they are lambs, but the wool lightens to oatmeal, light red, or a pale copper color. With the trend toward naturally colored wools, this fits the growing niche.

Fiber goats are almost always kept with their horns intact to help regulate temperature, and they have long, thick coats, as you would expect. Some breeds double in meat production depending on body type.
(Photo courtesy of R. M. Siegel)

Cashmere Goats. Cashmere goats are a type of goat, not a specific breed, so there are several breeds capable of producing the cashmere fibers. Many Cashmere goats are dual-purpose meat goats as well. Hardy, half-feral breeds like the Spanish goats have a natural cashmere fiber that can be enhanced with selective breeding. Most Cashmere, and indeed all fiber goats, are left with horns intact so they are easier to restrain and have better temperature regulation.

Lincoln Sheep. These sheep were originated from Leicester crosses and have been bred specifically for their long staple length and heavy fleeces. The wool fibers can be 8 to 15 inches long, and ewes can produce 15 pounds of wool a year! Lincoln sheep should not have wool below the eyes and are usually white, but also come in a variety of other colors, especially grey.

Romney Sheep. The Romney is sometimes considered a good dual-purpose sheep, as the sheep range between 150 to 275 pounds and have a tender meat quality. They produce an easily spun wool between 5 and 8 inches long. Ewes can produce 10 to 18 pounds of wool each year. White fleeces are common, but Romney will also be bred with natural colored and even variegated coats. The main focus of breeding with the Romney breed is uniformity in the crimp of the wool throughout the entire fleece.

OVER THE GARDEN FENCE

Take the time to see a few different breeds before selecting a goat or sheep breed. It's worth considering breeds that are available locally if you want purebred offspring so you have access to other breeders who can help you. Well-cared-for goats and sheep will live 8 to 12 years on average, although cases of them living 15 to 20 years are not unheard of. Make sure your choice is one that suits your needs and situation, because it's a longer-term commitment than some of the other livestock on the backyard farm.

Goat and Sheep Husbandry

Goats and sheep can be kept together without overly competing for resources because goats are more of browsers, while sheep are more grazers. There are some differences in the dietary needs between the two animals, which we will discuss in the following sections. Both can be offered high-quality hay as a free-choice supplement in a smaller backyard space.

The biggest challenge will be to ensure that their diet is balanced year-round without upsetting their digestive systems. Green grass and other forage, as well as hay, will make up the bulk of the diet with various feeds and supplements added in as needed. The basic needs are, of course, food, water, and trace minerals as needed. I'll discuss housing later in this chapter.

Care and Maintenance

Keeping goats and sheep can be more involved than keeping rabbits and chickens. I recommend finding a breeder in your local area to connect with to help you. In our case, we have a mentor we can turn to for help, advice, or just an expert eye. This kind of connection will be invaluable.

Here are a few things to keep in mind when caring for your goat or sheep.

Food. Food is obviously very important for your sheep and goats. The trickiest part for new home-steaders is understanding how your livestock's dietary needs change according to their age, workload, and whether they are currently breeding. A young goat that hasn't been bred yet won't need as much concentrated grain or feed as a doe that is carrying twins or milking nearly a gallon of milk per day. Sheep that are growing lustrous wool fleeces will need higher protein as well. The diet has to be monitored and adjusted for each member of your herd based on their individual needs.

This is one of the benefits of keeping livestock on a small scale. You know your animals individually, their quirks and their personality and their individual needs. I know that our Boer goat, S'mores, is always the first with her nose in the feed trough, so if she ever held back, I'd know something wasn't right. Both our mature does were bred this fall, and we already know both are carrying, so we've begun to increase their feed allotments until they kid (give birth) in another month. This personalized attention means your livestock will thrive and be healthier than they might be in a larger situation where that level of attention isn't possible.

We bring hay in as needed so our goats have constant access to roughage as they desire. We'll wrap this round bale with a cattle panel to keep the goats from wasting it and spoiling it.

The feed you provide consists of two basic types. *Roughage* is the high-fiber feed from browsing on bushes and shrubs, from eating grass, or from hay. *Concentrates* are supplementary and make up a smaller portion but are grains that are super high in energy, protein, and often in trace minerals needed for total health.

We offer grass hay free choice to our herd. They will barely touch it during the spring and summer when they are able to browse on the blackberries, honeysuckles, and other growing plants. But during the winter when growth slows down, they eat a lot more of the grass hay.

For feed grains you can supplement your herds with barley, corn, oats, and other grains. You can buy feed mixes specifically for goats and sheep also, which is a great way to start out. Be sure that when you are feeding commercial mixes you do not try to feed your goat mix to your sheep. Sheep should not have the copper minerals that are present in commercial goat feed mixes.

THORNY MATTERS

When you feed commercial feed mixes made for goats, you cannot give that same feed to your sheep. Sheep should not have copper in their mineral blocks, feed mixes, or food supplements, while goats do need trace amounts of copper. Feed your animals separately to avoid problems!

In general, you can assume that your goats and sheep will eat about 2 percent of their body weight in dry feed. So my 125-pound La Mancha doe will eat about 2½ pounds worth of hay and feed each day. Within that feed amount, at least 8 to 10 percent of her daily intake should be protein. In the last two months of her pregnancy, when the majority of growth occurs in the kids, her nutritional needs will increase and we'll replace part of her hay ration with alfalfa to help her keep her weight up.

Goats and sheep should never get overweight, because when they lay on fat it can be difficult to reverse the process. But you also don't want them to get too skinny, of course. Just petting your animal can give you a good idea of their condition—you should be able to feel their ribs when you pet them, and perhaps slightly see the ribs on thinner breeds like dairy goats, but you shouldn't be able to see the individual vertebra of the spine or clearly see the outline of the ribs through their coat. You also wouldn't want your goat or sheep too overweight to the point where you cannot feel their ribs.

One cool thing about feeding goats and sheep hay is that you can use "weedy" hay that horses and cows would turn their noses up at. We've actually secured some great deals by cutting our own hay on fields where the owners were trying to reclaim pasture that had overgrown. Lots of weeds, thistles, and brambles are mixed in with the grass, but the nutrition factor would actually be higher and provide the ruminants (animals that regurgitate partially digested food to chew it again) with the mixture of browsing materials that they really appreciate. Just be sure that it's not molded or old hay—it should still be fresh hay that is high quality.

Water. Your herds will need fresh, and most importantly, *clean* water. Goats and sheep won't drink soiled water so even if their bucket or trough looks full of water, they might not be drinking it. You'll need to replace the water if it's dirty so they will drink it. Never let the water run out, and be prepared for a doe who is about to kid or a milking doe to drink a gallon of water per day.

Goats need fresh, clean water available at all times. Remember that pregnant or milking goats and sheep will need more water each day.
(Photo courtesy of R. M. Siegel)

Grooming. Goats and sheep don't have to be groomed every day, but there are some things to be aware of, especially with fiber animals. During the seasonal shifts, especially when our herd members are losing their longer winter coats, we brush the goats on a daily basis. This gives us a chance to check their health and condition, and give them some relief as their sleeker summer coats come in. It's relaxing to them.

Right before kidding, you can clip the hair that might be growing near the udder to help the babies find their milk source after birth. This isn't necessary with all breeds. If you purchase your livestock from a reputable breeder, they will be able to show you which grooming methods are appropriate for the breeds you've fallen in love with. Bathing isn't necessary for goats unless there are lice problems in your herd. I've never seen a problem with lice, but our herd is smaller and not overcrowded, so that helps prevent issues.

Hoof trimming. Goats and sheep will usually need some hoof trimming unless your ground is rocky. How often depends on a lot of different factors, including each animal's personal development. A pair of trimmers and maybe a file are all you need to trim the hooves, and what you're trying to do is restore a balanced stance to the animal's foot. Any part of the hard hoof that isn't flush with the fleshy pad will need to be trimmed evenly with the pad of the foot so walking and standing is comfortable. If you, like us, adopt an older animal that had some neglect, you might find the hoof severely overgrown. The same tactics apply, though—trim away the excess hoof growth until the foot is properly aligned.

Housing

One of the benefits of keeping sheep and goats is that they are hardy animals with minimal housing requirements. While some facilities do provide a full barn situation, most ruminants prefer being outdoors and will do fine with a simple open-sided shed or shelter. It is vital to provide some shelter, however, to give your livestock shelter from temperature extremes, snow, rain, or even strong winds. You can see a simple open-door shed style plan in Appendix A.

The flooring must be dry, especially if you have an outdoor shelter. If the floor area of your shelter gets waterlogged you'll need to add gravel to build it up and create better drainage. We used wooden pallets on the floor and filled it in with dirt and gravel mixed to build the floor level up so it wouldn't hold water if it rained. On top of that goes a thick layer of straw and hay.

Concrete is easy to clean, but not comfortable for your animals, especially if they are probably going to be kidding or lambing in their shelters. If you must use concrete flooring, I recommend at least a foot of straw to cushion the floor. Dirt or gravel can be raked clean when you have a top cover of straw, and will be warmer and more comfortable than cold concrete.

Open shelters are fine in areas with no predators and mild climates. I've seen three-sided shelters that resemble cardboard boxes tipped on their side—only 4 feet tall so they are just big enough for the goat or sheep to get into. I prefer a taller shelter that is easier to clean out. Our goat shed is 10×8 and 6 feet tall, which is plenty tall enough for me to get into and clean everything up. It also leaves enough room for the milking stand to be inside the shed so we can milk the goats out of the elements on misty mornings.

This sheep is cozy in her warm shed with plenty of bedding. The wooden walls and sawdust will keep her warm, and most importantly, dry.
(Photo courtesy of Brian Boucheron)

Your shelter and feeding areas should be close enough to wherever you store the feed that daily chores aren't a hassle. We have an interesting arrangement in our backyard where the back wall of the storage shed doubles as the back wall of the goat shed. (See the quarter-acre illustration in Chapter 3.) Close proximity like that will make it extremely convenient to do your daily feedings.

Whatever housing arrangement you have should include a large-enough yard that your goats and sheep can have a comfortable amount of exercise each day. Our goat yard includes some interesting rock features that the goats like to climb on. If your yard doesn't have anything, provide wooden spools or large stumps for your goats to climb on.

We also have an additional yard that the goats can browse in on a regular basis. This allows them to change up their daily routine a bit, as well as providing them with fresh browse (brushy plants to nibble on). Having more than one yard or pasture area allows you to rotate where they are being kept and helps prevent (or decrease) parasite infestations.

Goats are natural climbers and will appreciate items in their exercise yard that allow them to practice their climbing skills.
(Photo courtesy of R. M. Siegel)

Signs and Prevention of Illness and Disease

Goats are usually quite hardy (sheep a little less so), but even the hardiest animals can sometimes get sick or infested with pests.

Parasites. Parasites are inevitable in your livestock, so you'll want to be on guard for when infestations become too severe. Parasites can be external, or living outside the animal like ticks, fleas, mites, or lice. Parasites can also be internal and are usually called "worms" and enter via the digestive tract. Both can create serious problems, of which malnutrition and anemia are only the start.

If you notice patchy areas of the animal's coat, or crusty wounds on their skin, you may have an external parasite problem. Sometimes mites will cause the animal to rub bald spots on their coat. Ticks, of course, are visible as they attach to the skin and swell when they feed. Internal parasites are harder to visibly diagnose, but if you notice your animals not gaining weight the way they should be, that's a good clue. You can have a stool sample analyzed as well, which is the best way to tell which specific parasites you should treat for. It's always best to only treat the specific parasites that are causing problems, and only treat when necessary, because resistant strains can become a problem otherwise.

CAEV (Caprine Arthritis Encephalitis Virus). This is a persistent problem in sheep and goat herds. The difficulty in eliminating it is that a goat can be a carrier of the virus without ever showing symptoms, thus not only passing it on to his or her offspring, but also infecting other goats and sheep they come in contact with. A blood test can tell you whether your stock is CAEV positive. If they are, you should make sure any kids or lambs born don't become infected as well. There is a growing movement to attempt to eliminate this virus from breeding stock completely, in much the same way that Coggins is being eliminated from horses.

Researchers believe that CAEV is transmitted through bodily fluids, blood, milk, and during breeding. If this is so, you should keep infected stock separate from uninfected stock. You would also quarantine any incoming animals and have them tested before introducing them into the herd. It is also okay to ask for test results of an outside buck or doe you are considering for breeding. In 1975, it was estimated that 80 percent of the goats in the United States were positive for CAEV, so you can see how attempting to eliminate it will take dedicated efforts on the part of breeders and owners. In 1999, several regions tested as low as 38 percent in the United States, so awareness is increasing. Similar efforts in Switzerland dropped the rate from 60 to 80 percent down to 1 percent.

Indigestion. Indigestion might not seem like a serious problem, but ruminants rely on a balance of healthy bacteria for their digestive problems. This can happen after antibiotics, or if the goat overeats harder to digest foods like grains. In many cases it's not contagious, but it's serious and can be fatal. Milk of magnesia can help detox and encourage movement in the rumen to get things moving again. Yogurt and probiotics will help restore healthy flora in the digestive tract. Waiting too long to treat this condition can kill, so be watchful and take action before it gets too serious.

A good breeder to purchase your initial stock, wise management in diet and care, and a great veterinarian are the three best ways to maintain a happy and healthy herd in your backyard farm.

Breeding Goats and Sheep

I love my baby goats. We joke that they are called "kids" for a reason—because they act like a bunch of little toddlers. They are curious, they love to explore, taste things, and play fight around the yard like a pack of preschoolers turned loose.

THORNY MATTERS

Before you breed your goats or sheep, you need to realize the level of care that babies involve—every day, even several times a day, if you end up with bottle-fed babies. If you cannot commit to that care, do not breed your animals.

The first step to healthy breeding is to start with animals that are in good shape. You'll never want to breed an animal that is underweight, overweight, or struggling with illness. We adopted a Boer goat who wasn't being cared for properly. She had been living with a domestic herd of deer, never handled, and poorly fed. She was wild, and quite a bit overweight as well because she had been fed kitchen scraps like white bread and other poor foods that should never be in your livestock's diet. So we kept her for almost a full year before breeding her to make sure she was back in good shape, that her temperament could be calmed down, and that she had her dietary needs met.

S'mores would not have been bred this fall if her attitude hadn't improved. She was never aggressive, only unused to being handled. Now she leads easily, comes to the fence first of all our goats, and has a great demeanor. Temperament should be in your mind before breeding.

You also want to have in mind what you are breeding for. Your desire for milk, meat, fiber, or showing your livestock should impact how you breed. In our herd, our primary concerns are milk and meat, so we aren't as concerned with pedigree, showability, or fiber quality. As a result, our herd is diverse with more than one breed, and we aren't concerned with color. We're using a big Nubian buck that had an excellent milker for a dam (thus passing on great milking genes) and has excellent size and build for crossing with our Boer doe for potential meat purposes.

Select a buck that will complement your breeding goals. He should compensate for any failings in your herd and should have a certification of good health from the veterinarian. Our small backyard farm is too small to keep a buck on site with our girls so we use a buck at a friend's house, taking our girls there in the fall for breeding. This arrangement works out well, giving us the chance to introduce new blood into our herd without the expense of keeping a buck that will only be bred with our two or three females.

You can also use artificial insemination, which means you purchase only the semen. The veterinarian will perform the insemination using what you've purchased without ever having your girls stand for a buck or ram. This is another viable option for a small-scale homesteader.

Most goats and sheep have periods where breeding can take place. Usually this happens in the fall and they come into heat every three to four weeks for a few months. With an experienced doe you will usually be able to tell when she's in heat by the signs she exhibits. Some of them are downright obnoxious about it!

OVER THE GARDEN FENCE

Signs that your doe or ewe is in heat include extra vocalizations, decreased appetite, wagging the tail (called *flagging*), seeking out a buck or standing against the fence near where he is penned, and decreased milk production. Ewes are less likely to show obvious signs of heat without the presence of a ram.

You can expose the doe to the buck you want her to breed with by introducing them to each other when the doe seems receptive. We usually move our girls over and allow them to breed at least two or three times confirmed. This last time we misjudged our Boer's heat cycle, so she ended up staying at the buck's house for three weeks. It was easier for us to leave her there than to upset her by transporting her back and forth multiple times.

If a sheep's long fleece gets in the way of breeding, you'll want to trim your ewe before breeding her. This could be as simple as a clipping around the tail end and udder areas for breeding and lambing ease. The ovulation rates of both does and ewes will be at their highest peak between the ages of 3 to 6 years old, which means breedings are more successful and will generally result in a larger number of babies.

Late gestation, the last month or so of the pregnancy, is when you'll want to provide the most nutritional support for your doe or ewe. If you allow the mother to lose too much body weight during this time, she may not be able to produce enough milk. And certainly, in a dairy animal, that would be especially unfortunate.

Lambing and kidding take place an average of 150 days after breeding. Mark your calendar for 147 through 155 days and start your watching and waiting period! This is so exciting, and every time my doe takes her time coming to the feed trough in the morning I rush out to see if there are babies in the shed. Sometimes they fool me, though, and have a knack for waiting until I've gone to the grocery store or a longer outing before settling into a birth.

An indicator of an approaching parturition (lambing or kidding time) has not been, in my experience, a swelling of the udder. I begin noticing a larger udder taking shape up to a month prior to birthing. A better indicator for me has been swelling or discharge of the vulva. Also, a softening of the ligaments around the hips and tail can make the tail seem to rise, while the abdomen seems to drop, almost separating from the spine. You can feel the softening of the ligaments under the tail and help to judge the mother's readiness.

Birth usually occurs within 12 hours of the onset of labor, and the pushing phase or second stage of labor happens within 3 hours. If you notice your doe or ewe seeming to struggle, and nothing seems to be happening after an hour or more, you may need to call in a vet. A poorly positioned baby may have difficulty coming out and may need assistance. This is something that many homesteaders and ranchers have done themselves for generations, but isn't something I recommend trying without experienced guidance your first time.

In most cases, your female will birth her babies without assistance. She should immediately begin cleaning the face and head of the baby. In the case of multiple births a doe or ewe may be distracted by the birth of the next kid or lamb and not finish cleaning the first baby. If this is the case you can use a towel to wipe the baby dry so it doesn't become chilled, but resist the temptation to do much more than that at first.

Within two hours of the delivery, the placenta or afterbirth should also be expelled. If you do not see the placenta come out, you will need to call a veterinarian because a retained placenta can cause a fatal infection. Note that each baby has its own placenta, so twin lambs means you should see two placentas birthed in the third stages of the labor. It is not uncommon for the mother to eat the placenta as a defense against predators being alerted to their presence, so if you do not observe the birth you may not actually see the placenta.

After birth has finished and the kid or lamb has been cleaned and dried, you should start to see signs that they are interested in nursing. An active lamb or kid will begin nursing within 30 minutes, while a slower-to-start baby may not begin nursing until an hour has past. Too much longer than that and you'll probably want to begin helping, especially in situations with more than one baby, if there's a chance that a weaker sibling will be pushed to the side. Kids and lambs should both consume about 10 percent of their body weight in their first milk, called colostrum, within the first day.

This baby goat is getting a final cleaning by her mother, although she's already up on her feet.

Colostrum is highly nutritious and beneficial for the kids and lambs. It has natural antibodies that are passed down from the mother to the babies. It's also really thick and rich to help stimulate their digestive systems right away.

Here is where two schools of management really diverge. Some herd owners will separate the kids and lambs right away, and raise them apart from their mothers. This can definitely increase the amount of milk available from the goats and sheep you are milking. The downside is the multiple times per day that you need to feed your kids. Others take a more natural herd approach and leave the babies with their mothers for feeding throughout the day. You can still get plenty of milk using this method if you begin separating the kids from their mothers at night, and milk the does or ewes first thing in the morning. You'll only get one milking per day because the kid or lamb will nurse the mother throughout the day, but this prevents you from having to bottle feed a large number of babies.

If you have more babies than teats (nipples) it can create a situation where one of the babies isn't getting enough milk. In this situation you have a couple different options. Bottle feeding is one of the most common. Another common method is to foster the baby out to another mother who maybe only has one lamb or kid and plenty of milk. In order to do this you'll have to mask the baby's scent until the mother accepts it, or tie the foster mother to allow the kids to nurse. I know goat owners who have good success rubbing vanilla extract under the doe's nose so she won't notice the baby's strange scent.

This baby goat is just a couple hours old but is already standing and has enjoyed its first meal.

This baby is only 3 days old. Bottle feeding your kids and lambs can increase milk supply, prevent the spread of CAEV, and make kids and lambs easier to handle as adults.
(Photo courtesy of SupernaturalNutrition.net)

In my experience, most of the kids wean naturally around 6 months of age. In 2011, the heat was so extreme that our doe didn't produce much milk through the full summer and weaned the baby a little sooner. We separate out the bucks at about 2 months of age because they have been known to breed does as young as 8 to 12 weeks. They are castrated by 12 weeks so that the buck-scent won't taint the meat. Female babies we keep unless we've reached my husband-imposed limit of four does for our backyard space.

Raising or Selling Your Yearly Offspring

Females that you are not planning to keep can be sold as young as 2 months old. If I'm planning to sell a doeling, I will usually not bother weaning them until the sale takes place. I'd hate to put them through the stress of weaning right before the added stress of a move. Your extra stock can sell for a wide variety of prices depending on the demand for what they have to offer.

Male offspring we castrate and raise to a butchering size at my in-laws' house. The arrangement we have with them is they provide the space, and we provide the feed. They tend the wethers until they are butchering size, and we split the meat that is produced. We've been able to add between 80 and 100 pounds of meat to the freezer each year in addition to having the doelings each year.

> **THORNY MATTERS**
>
> Castrating young males can be done with either banding or surgery. There are pros and cons to each method, and you should have experienced help before attempting either method yourself. Banding is bloodless but takes longer. Surgical castration does cause bleeding but the procedure happens a lot faster. Either way I recommend waiting a few weeks before castration to help prevent accidental damage to the urethra, which can cause a blockage of the urinary tract called urinary calculi.

Whatever babies you decide to keep will need to be vaccinated according to the schedule your veterinarian recommends. I also recommend handling them often to get them used to human contact. We teach them to lead at just a week old and have never had problems handling our adults who were trained young. We also get them accustomed to the motions of hoof trimming and grooming.

There is a limited market for castrated males that are trained to drive small carts, or used as pack animals for backpackers and hikers. Also check out local 4-H groups, farmers' markets, and feed stores for potential buyers. We've never had to list our goats in the newspaper or take them to the livestock auction, but both can be potential places to find buyers for your goats. Ideally you'll have developed a good relationship with the breeder that you purchased from and they can help advise you.

This goat cart race is an example of how castrated males can be used beyond their meat potential. Strong and agile wethers are also used as pack animals or friendly and hardy pets.
(Photo courtesy of Steve Swayne)

Milking a Goat or Sheep

We started out by purchasing an experienced milking goat already bred and about two months from kidding. This made it so much easier to get the hang of milking because we weren't training ourselves and training the doe at the same time. Since then we've learned to train our does when they are young to stand on the milk stand, and when they kid we're able to begin milking more easily.

Our greedy does don't raise too much of a fuss when we milk them because they have their heads stuck in a bucket of feed. We use milking time as their morning feed time and they quickly adjust to the routine. That's really the key—establishing the routine habit that the animal can rely on and grow confident in. Remember that any increases in grain should take place gradually, so it's wise to begin that process earlier rather than later to avoid digestive upsets.

If your milk station is not separated from the other goats, you'll need to tie them. We just tie the others to the fence while we rotate in whomever is milking. You want to have a set milking stand or station where you can have your equipment, and keep the area clean. Cleanliness applies to the goat, to you, and to all the equipment.

These milk stands are perfect for keeping the goats semi-restrained and easily feeding in the individual feed troughs during the milking process.
(Photo courtesy of Nicolás Boullosa)

Follow these steps for milking:

1. Settle the goat or sheep up on the milking stand and use a brush quickly to get rid of any loose hair and debris. Clean the udder and teat area completely and clean your hands.

2. Milk the goat or sheep into a nonreactive receptacle (we usually use a large glass jar with a wide mouth or 6-quart stainless-steel bucket) until you've emptied the udder.

3. We generally hand-milk our goats, and it only takes a few minutes to do so. Grasp the teat at the very top, where it meets the udder, with your thumb and forefinger in a circle. Squeeze those fingers gently and then push the milk that's now trapped in the teat out by gently squeezing your second, then your third finger in a fluid wavelike motion. Do not pull on the teat, as you can damage the ligaments that support the udder. Also take care not to twist the teat—the motion should be straight and even, and should simply squeeze the milk from the top to the bottom and out into the bucket.

4. Sanitize the teat immediately after milking as the teat canals will be open and susceptible to bacterial infections. You can purchase wipes specifically for this, or dip the teat in a disinfectant purchased for goats in a little disposable paper cup.

5. We bring the goat off the stand and tie her by the hay rack so she can eat a nibble of hay while we tend to the other goats.

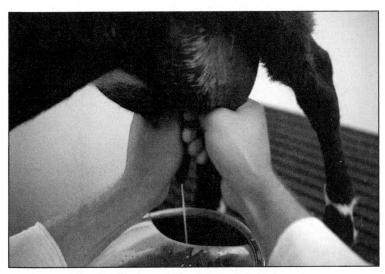

See how each hand grasps the teat firmly. Squeeze your fingers from top to bottom to squeeze the milk out.
(Photo courtesy of Brian Boucheron)

6. As soon as you've finished milking the first goat, take that milk in where it can be strained and cooled. We don't let the milk sit out in the goat shed for even the 15 minutes it would take to milk the second goat.

7. Strain the milk into a presterilized glass storage jar through a strainer lined with cheesecloth to get rid of any hairs or accidental debris that may have fallen into the milk bucket.

Here's an example of a large bucket of goat's milk being strained into the storage jar. Notice the strainer set inside the funnel.
(Photo courtesy of Nicolás Boullosa)

8. Place the glass storage jar immediately into the freezer to chill rapidly. After it's been in the freezer for 15 or 20 minutes (long enough to get cold but not long enough to begin to freeze), move it back to the fridge. Usually I can take 5 minutes to put the milk through the strainer and into the freezer, then go back to milk the second goat, and the milk from the first goat will be ready to move when I come in with the second batch.

Milk will keep about four or five days in the fridge, and if there is any leftover at that time we move it out to the shed to turn into clabbers. This fermented milk is a delicious and high-protein treat for chickens, hogs, our dogs, or the barn cat.

MILKING TOOLS

Milking tools are especially handy if your hands are too large or if you have arthritis. One hand-milking tool that is very reasonably priced is a simple vacuum pump-type system that you just squeeze and collect the milk into a jar that's attached. Fancier milking machines are motorized and can sometimes milk more than one animal at a time. Obviously, the fancier you get, the more money you'd be investing and the less likely it is to be a good investment for the average backyard farmer. Check your local feed store or online farm supplier for options.

Shearing Sheep or Goats

If you're growing a fiber breed of sheep or goat, you'll want to eventually harvest your fiber. Cashmere goats are not shorn, but much like Angora rabbits (see Chapter 13), they are combed or plucked to get the soft, spinnable fibers. Sheep with wool fleeces, and Angora goats (which produce mohair fiber) can both be shorn. Shearing sheep or goats is tough work, but for the fiber enthusiast it is also an exciting time.

Angora goats usually get two shearings per year, once before breeding and once before kidding. This allows the fleece to stay as clean as possible. There's a balance in the timing because you want to be harvesting the longest fleece possible and wait for it to grow out, but you also need to harvest before it starts getting dirty and matted. Because this shearing will remove the entire fleece, your goats will need extra protection from bad weather postshearing. Add extra layers of straw to their shed and make sure they do not get damp. You may also need to fit each with individual coats temporarily until their own coats begin to regrow.

It may look uncomfortable for the sheep to be flipped over during shearing, but notice the other sheep waiting its turn. They are ready to lose their coats for the summer!
(Photo courtesy of Brian Boucheron)

Sheep are sheared in much the same way. You can shear them by hand with hand-clippers or use electric clippers. Either method should be done with care and can be tough to accomplish. It helps to begin working with your sheep when they are young so they become accustomed to you and being handled. Often you will want to time your shearings for early summer.

Sit the sheep up or roll her over and clip the stomach area. If possible, hold her steady with your knees so you can use both hands for better control. Take care around the teats and cut as close to the skin as possible without nicking the tender skin. Continue clipping each row further to one side until you cut all the way down the sides and almost to the back. Then work the fleece from the other direction until you meet your cuts in the middle. Shear the tail end and back of the legs carefully, and turn the sheep loose. With practice you'll be left with the full fleece in one large cutting!

With any fleece, whether from your goat or sheep, you'll want to remove any sections that are matted, dirty, or soiled by feces. Then your fleece can be rolled up and sold in the grease (unwashed) or washed and sold at its washed weight. Fleeces will lose some weight after they have been washed and then dried because the oils produced naturally by the animal will be washed away, leaving the fiber fleece lighter than it was originally.

This fleece has been washed and is spread out to dry. Notice how well it held its structure during washing to preserve the workability of the wool.
(Photo courtesy of Emma Jane Hogbin)

To wash fleeces, you can use either a washing machine or a large utility sink or tub. Use a mesh bag to help hold the fleece in place and to prevent felting (unusable matting) of the fleece.

1. Let the washing machine or sink fill with hot water and then let the fleece soak completely submerged in the hot water.

2. Drain the water and either spin the fleece in the spin cycle or press it dry between two towels (do not twist or wring).

3. Refill the washer or sink one more time with a mild dish-washing detergent so the soap is agitated before adding the fiber.

4. When the water is soapy, submerge the fleece and let it soak for at least 30 minutes more to clean thoroughly. Keep the water hot or the grease will stick onto the fleece again.

5. Rinse thoroughly in hot water again just as you did in the first step. If you have to repeat the soapy stage for a particularly dirty or greasy fleece you can, but do not swish the fleece around in the water or it will fuzz up and be ruined.

6. Let the fiber dry thoroughly while lying flat on a screen. Keep it out of direct sunlight but provide plenty of ventilation and air flow. Do not put a fan on it because you'll fuzz out your fleece.

When it's completely dry, you can roll it and sell it at a higher price per ounce or per pound with a clean and workable fleece!

Beekeeping in the Backyard  15

It's one of the fastest-growing city homesteading trends in the nation—city beekeeping. The year beekeeping was legalized in New York City, the local beekeeping club had more than 200 members. Beekeepers have been banding together to change restrictive legislation with good success in recent years.

Thankfully, the trend toward self-sufficiency is being supported by local regulations recently with more and more cities loosening restrictions. Some cities are highly supportive, such as Chicago with its rooftop beehives on its city hall building. One of the best resources for finding out the legalities of your particular city or county is to check with your state apiarist. Go to bit.ly/BYFBeeStates for the contact information for your state.

Getting to Know the Bee World

Beekeeping comes with its own lingo that you'll want to become familiar with. Here are some common terms you should know:

- **Apiarist**—A beekeeper.
- **Apiary**—The beeyard where the beehives are kept.
- **Balling the queen**—An aggressive behavior by worker bees who crowd around and attack a queen they perceive as foreign.
- **Beeswax**—The glandular excretion honey bees produce naturally and use to build honeycombs for raising young, storing pollen, and of course holding honey.
- **Brood**—The baby bees, as eggs, larvae, and pupae stages.
- **Brood chamber** or **brood nest**—The portion of the hive where brood is raised.
- **Cappings**—The tops of the capped honeycomb cells. Both brood and honey can be capped.
- **Cell**—The individual, single hexagon cup within the honeycomb.
- **Colony** or **hive**—A group of bees that live together in a single hive unit.
- **Drone**—The male honey bee.

- ⚶ **Honey**—The sugary liquid produced by bees and stored in the combs. Highly concentrated source of energy that bees use to survive the winter months.

- ⚶ **Nuc** or **nucleus**—A miniature hive box with two to five frames used to split hives. Often a nuc can be purchased when beginning a new apiary.

- ⚶ **Queen**—The female in the hive who is capable of laying eggs and producing offspring. A virgin queen has not yet bred with a drone. After a mating flight a queen is a fertile queen.

- ⚶ **Swarm**—Unlike in horror movies, a swarm of bees is a docile, if scary-looking occurrence. Swarms happen when a hive splits and a large portion of the workers leave the hive with a newborn queen to start a new hive. It is the bees' natural method of creating new hives.

- ⚶ **Worker bee**—The sterile females that do the majority of work around the hive. They build the comb, tend the brood, and gather nectar and pollen.

Starting Your Apiary

Starting your beehive is an exciting time. I recommend really thinking about where you place the hive, especially in the city where you have neighbors closer to you. There are ways to set up and manage your hive to be more "neighbor friendly." Bees don't like to move after they've established their new home, so plan your location carefully and provide what your bees need from the start.

A 6-foot fence or line of shrubs about 10 feet or more in front of the hive's opening will force the bees to fly out and then up. This keeps the bees from choosing a flight path right at someone's eye level or along a busy sidewalk. Encouraging your bees to fly up out of the way will go a long way toward preventing accidental stings.

Bees need a plentiful source of water. Often after they settle on a water source they will continue to use that water source even if you put one closer to the hive later. Provide water for your bees near their hive right from the very beginning so they don't choose your neighbor's water feature or wading pool for their water source. You can actually get a waterer that fits in the entrance of the hive specifically for this purpose. Or add a little bird bath within about 20 feet or so.

Bee colonies should be placed in a sunny location that is sheltered from wind and extreme winter weather. It's also helpful, but not necessary, to elevate your hives so they aren't sitting directly on the ground. Simple cinderblocks will work for this purpose.

Equipment Needed

There are a few things you'll need before you can keep bees, and unfortunately, buying these items brand new can be a bit of an investment. If you're fortunate enough to find inexpensive, used equipment, you'll want to be sure to sanitize it so you don't spread diseases or pests to your new hive.

This city beehive is situated so the entrance is away from the neighbor's house, which makes for happier neighbors. The hive is also situated nicely on cinderblocks for a stable base. You can see the bees on the entrance board buzzing in and out away from the neighbor's house. With the hive entrance slightly lower than the back of the hive, rainwater cannot pool inside.
(Photo courtesy of Nicolás Boullosa)

The most important, and usually the most expensive, item you'll need to purchase are the hive boxes themselves (called *supers*). The most common hive setup has boxes with no tops or bottoms that hold individual frames within them. These frame boxes can be stacked on top of each other to increase the size of the colony.

The most commonly available designs are 8-frame boxes or 10-frame boxes. We use 10-frame boxes because that's what was available to us. We were given access to a set of old hives that had been in a shed for probably 20 years. They needed a lot of work to clean them up, repair the frames, and add new foundation comb, but we didn't have to spend as much money on the initial setup, which was a huge boon. Obviously it's easier to start with new frames from a beekeeping supply company.

The hive boxes come in three sizes, a deep super or brood super, a medium honey super, and a shallow honey super. The deep supers are used as the bottom box or bottom two boxes and are left for the bees to raise brood, and store pollen and honey to feed themselves. The other supers are used for the honey collection. If you use a queen excluder—a screen that has holes big enough for worker bees to crawl through, but not big enough for the queen to crawl through—the queen will only be able to stay on the bottom, and only the workers will be able to get to the top boxes where the excess honey will be produced for you. If the queen cannot reach the top boxes, she can't lay eggs there. Then when you harvest the honey from those top boxes later, you'll be able to harvest only honey and comb without worrying about whether eggs have been laid there.

This is an eight-frame hive box with shallow frames. These boxes will be easier to handle when they are filled with honey because they will be lighter.
(Photo courtesy of Nicolás Boullosa)

Note the queen excluder screen in front of the hive. These holes are too small for the queen's larger body but big enough to allow the worker bees to access any supers placed above the excluder. As this hive grows, additional supers will be added above the excluder, keeping the queen from laying eggs on frames with honey.
(Photo courtesy of Meg Allison Zaletel)

Honey supers are usually not as deep as the brood boxes, because frames filled with honey are extremely heavy. By making the boxes shallower they will be easier for you to lift. The size doesn't limit what the bees are capable of producing, because when they begin to fill up you'll simply add another box to the hive.

You'll want some protection starting out, especially a bee veil to protect your face. Veils are usually considered the minimum protection to wear when working your hive, and for many experienced beekeepers it is all they wear. Depending on your comfort level, you can wear a full protective suit. Gloves can often create more problems than they are worth because they make it easier to accidentally squash or mishandle the bees. However, they are handy when you're starting out and a lot will depend on your comfort level.

I highly recommend joining a beekeeping club before purchasing your bees. Find some of the more experienced beekeepers and inquire when they will be working their hives, and visit them during that time. You'll get a chance to observe what a full hive looks like, and your beekeeping mentor can explain what you should be looking for whenever you open your hive. You can also practice handling supers and frames and see what level of protection you feel most comfortable with. I've never owned a full suit, but I absolutely cannot bring myself to grip the frames without wearing a pair of gloves.

THORNY MATTERS

Bees release an alarm pheromone when they are killed or when they sting something, which can trigger more aggressive behavior from the remaining bees. Take care not to crush or pinch bees while working the hive and you'll be less likely to get stung.

One of the best tools I've ever purchased was a frame grip. It acts like a pair of pliers to grasp the frame and help you lift it out. It minimizes your contact with the frame so it'll be less likely to kill your bees when lifting the frames out. It also keeps your fingers from grasping directly on the frame so you're less likely to get stung by trapping a bee under your fingers.

A hive tool is a crowbar type of instrument specifically designed to help lift the lid, and separate the hive frames so you can lift them out and examine them. Bees will stick the frames together with a sticky substance called *propolis* to secure them. Just pop them loose to lift them out.

You'll also want a smoker. A smoker will create a cool, persistent stream of smoke that will help calm the bees while you're working with them. Often when you open the lid of the hive, the bees will pour out of the top. When you're ready to replace the top, you should use the smoker to drive the bees back down into the hive so you minimize casualties to your worker bee force while working the hive. It's impossible to check the hive without accidentally squishing a few of the worker bees, but you obviously want to minimize that as much as possible.

When you lift the top off the hive, the bees will naturally pool up to the top of the frames. Smoking helps drive them back down into the hive so you can work the frames.
(Photo courtesy of Nicolás Boullosa)

There are many other items you can purchase, but these are the bare essentials that you'll want to have on hand. Of course the longer you keep bees, and the bigger your hives become, the more tools you'll be able to invest in and get your money's worth.

Acquiring a Hive

There are a few different ways of getting your first beehive after you have your hive boxes set up and all your equipment on hand. You can install a captured swarm. You can purchase a bee package. You can purchase a nuc hive. You can purchase a full colony. These options all have their various pros and cons, and represent different price points for getting started.

Catching a swarm of bees can be either ridiculously easy or extremely difficult, depending on the location of the bee swarm. Ideally a swarm would be on a low-hanging branch or other easily accessible location. If the swarm is in a difficult or dangerous to reach place, the attempt could harm you and wouldn't be worthwhile. A 5-gallon bucket or sturdy cardboard box may be the only container you need for catching the swarm!

You have to understand that a swarm of bees is extremely passive. They do not try to sting anything; they are mostly interested in keeping the queen safe, which is why they will often appear as a giant, living ball of bees huddled around the end of a tree branch or lamppost. The queen is somewhere in the center of the ball and if you knock her into the bucket, the worker bees will follow and you'll have a large, young bee colony ready to install into your hive.

If you've joined a local beekeeping club, ask one of the more experienced members to help you with your first swarm. They are usually very willing to help, often for free or for a nominal fee, which is what makes installing a swarm one of the least expensive options for starting a new hive. Take the swarm to your hive and tip the bucket so it's near the opening of the hive and they are likely to just crawl inside. Otherwise take the top off the lid of the hive and pour the bees onto the top of the hive. As long as the queen goes into the hive, the worker bees will follow.

Installing a package of bees is much the same way. The only difference is you order the bees and they are shipped to you with queen and workers ready to go. Your bees will arrive in a package where all the worker bees are inside and the queen is in a separate compartment of her own. This is by far the most common method because it's more reasonably priced than purchasing a new hive or nuc, but it's easier and more reliable than trying to find a swarm. You can usually find a package of bees for under $100.

A package of bees ready to install into their new hive. These are usually purchased by the pound.
(Photo courtesy of Meg Allison Zaletel)

During the shipping process the worker bees begin to become accustomed to the queen bee and will start to accept her as their queen. The queen's cage is usually plugged by a sugary, almost candylike plug that the worker bees will eventually chew through. Don't release the queen out of her cage yet—let the bees do that on their own. By the time that happens, the queen's scent will have been dispersed throughout the hive's workers and she'll be the accepted queen.

This is the queen cage, where the queen is protected from being killed by the worker bees until they've accepted her.
(Photo courtesy of Meg Allison Zaletel)

Sometimes experienced beekeepers will have packages available for sale as they split their largest hives to prevent swarming, so check with your local beekeepers before ordering a package through the mail. The local stock will be less stressed by a long shipping time, and will be well adjusted to your local climate! The downside to purchasing a split from a local beekeeper is that you risk buying a package of bees that is infected with mites or an illness.

Nucs are one of my favorite ways to establish a new hive. Instead of purchasing only the bees themselves, you are purchasing a few frames of filled goodness to go along with the bees. It's basically a mini hive box with anywhere from three to five frames. We started with five frames in ours and got three frames of brood, one frame of honey, and one frame of pollen. Plus of course lots of bees and a queen are already established in this mini colony.

The benefits of beginning with a nuc is that you have brood that is ready to hatch in various stages of development, which can grow your hive a little faster. Eggs usually take 21 days to go from brand-new egg to fully mature worker bee. By giving your bees a jump on increasing the number of workers right away, your population won't suffer as big a decline before the newly hatched workers can help boost the workforce.

The downside is that nucs can be more expensive to purchase than just a package of bees, and you will need to make sure that the frames you are purchasing will fit in your hive box. Some nucs come with deep body frames, and others come with medium body frames, so be sure you know what you're buying! Nucs cost around $120 to $150 and are available from local beekeepers or through mail-order apiary supply stores.

The last option for starting your beekeeping hobby is to purchase a mature and thriving colony. This of course gives you an immediate return on your investment, with a full honey harvest the very first year. However, it is the most expensive option. You run the risk of buying outdated or faulty equipment, but on the other hand if you already have to purchase hive boxes and frames, your cost for an established colony might not be that much more than the price of all new boxes and frames plus the bee package.

Regardless of how you get the bees, you'll need to have your equipment set up and ready when the bees arrive. And of course, the same principles apply to bees as to all livestock. The initial purchase is only the first, and often the easiest step. After getting your bees home, the work begins.

Keeping Bees—A Year-Round Guide

Let's look at how you'd go about working your hive throughout the year. I want to stress, though, that this is an overview of what to expect throughout the year in a general sense. I strongly urge new beekeepers to connect with a local beekeeping group. You'll have the wisdom of expert bee-keepers who know the specifics of your local region and their advice can literally save you hundreds of dollars. You can ask your State Apiarist for contact information about a group in your local area. Find the State Apriarist in your state at bit.ly/BYFBeeStates.

Start in the early spring with your newly acquired hive—the time when most new hives should be purchased. Spring is the time when many plants are blooming, providing steady sources of pollen and nectar to feed your bees. This also gives the newer, less stable colonies a chance to grow stronger before the trying winter sets in.

Let's say you get your package of bees installed by the end of March or April. Your queen will be released from her cage by the third day (rarely will you need to release her yourself on the third day) so you'd want to check the hive the third day to make sure. Let the hive be for another couple of days after you release the queen so she can settle in. When you check the hive at the end of that first week, you'll hopefully see brand-new eggs in brand-new combs. Those hundreds of tiny little eggs are the sign that your queen has settled in and is doing her job—by building up the workforce to strengthen the hive.

Check the hive again three weeks later and you should see the first of those new eggs beginning to hatch as the first of your worker bees emerge. Congratulations, you are a bee parent!

Installing a new package of bees is best done after the weather warms up, but as early in the spring as possible to give them time to build the colony strong enough for winter. Just pour the bees into the hive and they will settle into the frames there. A captured swarm would install the same way.
(Photo courtesy of Meg Allison Zaletel)

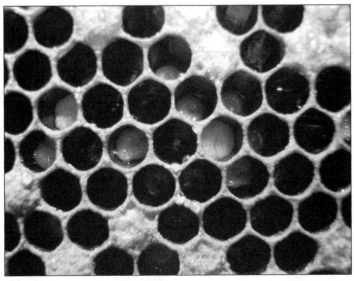

Bee eggs and larvae growing in the hive. This is a good sign that the queen is alive and laying.
(Photo courtesy of Emma Jane Hogbin)

You'll want to continue feeding the hive until there is a good *nectar flow*. After you have new worker bees hatching out, and there are plants in your region providing nectar for the hive, they will decrease the amount they are feeding from the artificial feeder. This can happen in around two to six weeks, depending on how quickly your hive takes off.

DEFINITION

Apiarists call a **nectar flow** (sometimes called *honey flow*) the time when there is a surplus of nectar coming into the hive and the bees are rapidly expanding their colony, food stores, and brood.

When you see that your bees have used six frames in a box, you'll want to add another super. Check often enough during a nectar flow that you can add more supers as needed. This flow will continue all through spring and early summer.

When July and August hit, there are often a few weeks of famine, often called *dearth*. It's common in many areas of the United States and is a good time to decrease the number of empty frames available to the bees above the queen excluder. The bees will switch out their storage habits and begin exchanging some of the brood frames in the deep supers to nectar and pollen storage for surviving the famine. This is totally okay.

Think about it this way—the more there is to harvest, the more worker bees (new brood) you want available to work the nectar flow at its peak. After the nectar flow begins to decline, you want fewer mouths to feed because not only is there less to harvest, but now there are fewer bees to feed heading into the winter. The queen lays fewer eggs as the winter approaches, so as workers born in the summer months begin to die, they are not replaced at the same levels and the overall population level of the hive drops slightly over the winter.

One of the best ways to minimize the number of supers in the hive during this transitional period is to harvest the frames of honey that are totally capped, and leave the unfinished frames. This allows your bees plenty of space to continue harvesting and producing more honey for you. But at the same time it encourages them to settle into the task of filling each cell left in the super and not leave any gaps so your harvest will be maximized.

This late-summer harvest is when you get to harvest the surplus. Not all the honey, by any means. Remember when I mentioned leaving one or two boxes as the hive body? Those boxes are for the bees. The supers on top with the excess honey (if you get any excess the first year) is what you get.

When you reach October, you'll want to have at least 70 pounds of honey in the hive for your colony to survive the winter. We've lost a hive to winter before and it had nothing to do with the cold. A well-fed beehive is capable of surviving extremely cold temperatures if they have enough food. Our drought conditions were so severe in 2011 that our brand-new hive never had a chance to become well established.

Bees will fill individual cells with honey, and when it is ready, they cap it with beeswax to store it. This honey frame is not fully capped. Notice the white wax caps in the upper-left corner and across the top. When the entire frame is capped, it is ready for harvest.
(Photo courtesy of Emma Jane Hogbin)

You may need to feed your bees during this time so they have enough food stores to make it through the winter in good shape. This is only necessary if they don't have enough food collected on their own. One of the simplest mixtures is a 1 part water to 1 part sugar mix. But if your bees are in bad shape you may want to increase the amount of sugar so it's a much thicker, syrupy mix of 1 pound sugar to 1 pint of water.

There are several types of feeders for feeding bees, including entrance feeders, simple handmade jar and platter feeders, and zipper-lock storage baggie feeders (which, because it's placed inside the hive, is less likely to attract robbers). I don't recommend the division board feeders because they can drown your bees and if you are trying to help a colony through a rough patch, the last thing you want to do is kill them off.

THORNY MATTERS

If you look near the entrance of a beehive, you'll see guard bees there protecting the hive from intruders. Bees are attracted to sources of nectar, including the honey collected by other hives. A weak hive is susceptible to robber bees—workers from other honeybee hives that will come in to steal the nectar stores from your bees. When your hive is new, and during the winter when bees are clustered for warmth, an entrance reducer will make the wide entrance small enough to be more easily guarded as well as prevent mice from coming into the hive.

Late winter is the time you need to be ready to inspect your colony again. The first nice warm day in January, be ready to check it out. Is your queen still alive? Do your bees have food and brood, and are there any new eggs being laid? If you see any potential problems, check with your local mentor for tips on how to remedy the situation.

Learning to recognize and find the queen is important for a beekeeper. You'll want to see her or see signs that she's been laying recently. The queen in this photo is almost dead center with a much longer torso than her daughters. This enables her to reach the center of the cells to lay her eggs.
(Photo courtesy of Emma Jane Hogbin)

Toward the end of January and beginning of February, you want your queen to be laying eggs so your new workers will be ready a month later to gather the nectar during flow. This means you may need to feed the 1:1 syrup again now to stimulate your bees into waking up for the year. This three-week jump on the coming spring season can make a big difference in honey production come March!

The Honey Harvest

The fun part! That sweet, sugary goodness that is the result of thousands of work hours on the part of your bees is such an amazing miracle. Sometimes the abundance can be overwhelming when you hit a good year with a well-established hive and suddenly have 100 pounds of honey in your kitchen.

Tools for Honey Extraction

Most beekeepers end up investing in a honey extractor. These machines come in all manner of set-ups, but most work the same basic way. You place the frames full of honey inside and the machine spins using centrifugal force to release the honey out of the combs. Note that the frames have to be uncapped before the honey can be extracted this way, so you'll need a decapping knife or hot knife to cut through the top coverings of the honey cells.

Many of these machines will spin several frames at a time. This leads to some creative bartering amongst beekeepers and joint harvest parties where you might be able to take your honey frames to someone else's machine. Some beekeepers will harvest by hand methods rather than risk mixing their honey with another beekeeper's—especially if they've been trying to use organic methods.

After the honey has been extracted, you'll return the frames to the hive temporarily so your bees can collect the beeswax cappings and clean any remaining honey from the frames. Be sure you don't leave these still-sticky frames out in the open or you'll encourage robber bees to come and attack your hive! Keep them covered until you can return them to the hive for a day and let the bees clean them up.

One of the big benefits of spinning the honey out of the frame is that you preserve the individual cells of the honeycomb. This gives the bees a head start in filling those frames the following season, because they won't have to build the comb and then fill it—they can get right to work filling all the cells.

When the honey has been extracted, it will need to be filtered. Again there are several methods for accomplishing the same thing—straining the honey through a fine mesh or fabric filter that will remove any wayward debris, dirt or pollen grains, or leftover wax that came loose during extraction. Allow the honey to settle so all the air bubbles are at the top, and then fill the presterilized honey bottles or jars with your bounty!

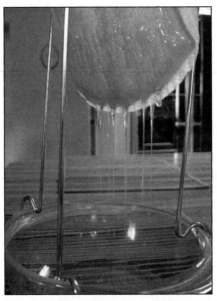

These backyard farmers used a jelly strainer to strain their honey harvest.
(Photo courtesy of Meg Allison Zaletel)

Comb Honey

If you choose to harvest comb honey, sometimes called honey in the comb or chunk honey, you won't have to run your frames through an extractor because you'll be selling both the valuable beeswax combs and the honey within as is. It also has that traditional charm that is so appealing. The most popular form of selling comb honey is a combination of extracted honey with a large chunk of comb in the jar surrounded by extracted honey to fill in all the gaps.

When you are harvesting for comb honey, you have to decide that ahead of time because it will change the type of foundation or central backing you use in your hive's frames. If you use wire-embedded foundation you can't produce true comb honey because of the wires. Ultimately you want high-quality "virgin" beeswax that is brand new, filled with honey, and capped with thin, delicate cappings so your comb honey is both attractive looking and highly edible.

You'll cut the comb out of the center of the frame (instead of spinning out the honey and leaving the comb in the frame) to harvest your comb honey. Trim away the rough edges so all that's left is fully capped, uniform-looking comb. Cut chunks according to the size of the jaw, using the largest dimensions that the jar will hold. This is a lot easier with a wide-mouth jar, of course. Then fill the jar with extracted honey and you'll have a gorgeous, highly saleable jar of honey straight from your own backyard.

Straining the Honey

To extract honey the cheap, but very messy way, you will actually cut out the comb and instead of chunking it up for comb honey, you will crush it to release the honey. You'll uncap and crush the honeycombs so you have a giant mash of honeycomb and honey. Then you'll have to strain the honey to separate it from the comb pieces just as you would strain extracted honey.

After you've harvested the honey, at the minimum you'll have the cappings left over. And if you've used the crush and strain method you'll have a whole mess of beeswax left. For many crafters and home DIYers, this is a bonus!

Capping and new wax is the best for crafting purposes, as older wax will have more propolis and other contaminants in it and will be a darker color instead of the light yellow color of new beeswax. When you've strained all the honey out of the cappings and beeswax you'll be ready to melt it down. You may need to wash the cappings in order to get all the honey out. This is a messy process, but fun.

This leftover wax and cappings can be rendered into useable beeswax.
(Photo courtesy of Meg Allison Zaletel)

To render the wax, or melt it down, you'll need a heat source like a gas stove or electric hot plate. You want to be careful to use indirect heat such as a double boiler to heat the wax, because beeswax is flammable. If it gets too hot or comes into direct contact with the flame it can catch fire, so use caution. After you've melted the wax in your double boiler, you'll want to strain it through some kind of filter to remove the debris from the wax.

Some people will just ladle out the wax, being careful not to scoop the wax from the bottom of the pan because all the debris will settle to the bottom. That's an option, too, but usually makes it harder to get a consistent wax rendering. Pour the wax through the strainer (or scoop it) into temporary containers such as yogurt cups or butter tubs until it cools. Now your rendered beeswax is ready for sale or craft use. Chapter 25 has a couple ideas on ways to use it.

Enjoying the Bounty

 **4**

It's not enough to just grow and produce the useable. As a backyard farmer taking your place in sustainable living efforts, you want to be able to preserve and partake. These chapters show you how.

Learn to enjoy the richness of the flavors available to you in each season. Bask in the uniqueness that each turn of the earth has to offer. Eating in season is an art of appreciating the finite, a far cry from attitude that says "Give me what I want, when I want it, no matter the cost."

Preserving what is available for future months is part of that process. Long-term storage that prepares for times when something isn't available is a wisdom of the ages that needs to be revived in today's society. Learn how to can, smoke, dry, store, and lay up for the future, and enjoy the luxurious bounty available to you as a result.

A Seasonal Guide to Managing Your Harvest

Have you ever had neighbors sneak zucchini onto your front porch and run away hoping you'll adopt their overly productive garden fruits? It totally happens on the backyard farm as seasonal bounties ebb and flow. One month you have an influx of tomatoes, the next month your meat chickens are ready to put in the freezer, and the next month you have more winter squash than you know what to do with. But nothing should go to waste on a backyard farm, and you can preserve what's available now for the times when it isn't available later.

Learning to appreciate what is available in season during the peak of its availability will make best use of what you're growing and producing. Not only that, but you'll be eating more nutritiously as well because the vitamins and minerals in your food decrease with storing and preserving. And of course, eating locally (or self) grown foods while they are in season will save you money as well.

There are so many reasons to take the challenge to eat more seasonally. It's a shift of focus for most of us, but an important one for our Earth and our self-sufficiency.

Eating Fresh in Spring

As winter chill fades away, the world awakens with spring's promise. In the garden, spring means you'll be harvesting anything left from your fall garden and late-winter garden (Chapter 7). Swiss chard, peas, lettuces and cabbages of all kinds (Chapter 8), and soon fresh asparagus (Chapter 10) will be ready for harvesting.

On the livestock side of things, you'll be on kid watch for your goats, or lamb watch with your sheep (Chapter 14). You'll be getting ready for this year's baby chicks (Chapter 12) and checking your hives for signs of nectar flow (Chapter 15).

Gathering fresh spring greens, like spinach or chard, along with those amazing first eggs as your hens bask in the lengthening days, is indescribable. And everything works in tandem. It's time to begin sowing seeds, both indoors and out, and your goats, chickens, or rabbits appreciate the thinnings from extra seedlings. Meanwhile, your spring cleaning deep clean of coops, sheds, and cages provide compost materials that will feed your garden in the coming year.

Spinach Salad with Tangy Dressing

The sweet and tangy dressing makes this the perfect easy-to-prepare salad for looking like a superstar in front of company, or just to treat your family to an amazing dish.

Yield:	Prep time:	Serving size:
4 cups	20 minutes	½ cup

¼ cup granulated sugar

1 tsp. salt

1 tsp. dry mustard

1 tsp. poppy seeds

1 TB. onion juice

⅓ cup apple cider vinegar

1 cup vegetable or olive oil

¾ cup cottage cheese or soft cheese (such as soft goat's cheese)

2 lb. fresh spinach leaves, rinsed and patted dry

2 or 3 hard-boiled eggs, chopped

8 strips bacon, cooked until crispy, drained, and crumbled

¾ cup fresh mushrooms, sliced

1. In a small bowl, combine sugar, salt, dry mustard, poppy seeds, onion juice, apple cider vinegar, vegetable oil, and cottage cheese.

2. Divide spinach leaves, eggs, bacon, and mushrooms among four plates and drizzle dressing over salad.

3. Serve immediately.

Spinach won't last much longer than the first days that peak in the 70s, so as soon as temperatures begin to be consistently warm, harvest the rest of your spinach and freeze it after blanching (plunging into boiling water for two minutes). Until then, enjoy the delicious and unique flavor spinach adds to your fresh salads! When harvested straight from the garden, there is no comparison.

Swiss chard and asparagus I like best cooked in similar fashions. Melt a little dab of soft butter in a skillet and sauté with minced garlic. Swiss chard should be prepared by cutting the stems in bite-size lengths and slicing the greens.

Boil the Swiss chard until it's tender (about one to three minutes) and then sauté in the butter and garlic mixture. Top with dried basil and pepper to taste. It makes an excellent side dish that can't be beat in the spring.

One of my favorite cool-season vegetables, Swiss chard, is easy to cook and hardy in the fall garden.

Radishes are one of those super-fast growing, early spring root crops that no one ever knows exactly what to do with. But everyone grows them. Especially if you have kids, because radishes are so easy to grow, it's a fun one for the little guys to play with. But what to do with them?

Radishes and Egg Salad

Perfect for a picnic, this dish is a good spring variation on traditional potato or egg salad. The peppery flavor of the radishes mixes perfectly with the sour cream.

Yield:	Prep time:	Serving size:
4 cups	20 minutes	½ cup

1 lb. radishes (red or whatever color you can find)

3 hard-boiled eggs, chopped

1 bunch scallions (about 6)

½ cup sour cream

¼ tsp. salt, or to taste

Dash of black pepper

¼ tsp. dried dill or fresh dill weed, finely chopped

1. Wash and trim roots and stems from radishes. Cut each radish in half from stem to root, then cut each half into slices as thin as possible.

2. In a large bowl, add eggs to radishes.

3. Wash and trim roots from scallions and use the green tops as well as the white parts. Cut into slices between ¼ and ½ inch wide. Add to the bowl, along with sour cream, and sprinkle salt, pepper, and dill evenly over all.

4. Stir with a large spoon or rubber scraper until well mixed. Place in serving dish and garnish with more dill if desired. Serve immediately, or cover and chill. Refrigerate any leftovers, and use within a couple days.

(Recipe courtesy of Home-Ec101.com)

Summer's Feast on the Kitchen Table

Summer is a time of great activity. It's a good thing the days are longer because otherwise you wouldn't be able to get everything done! Not only are things crazy busy with your garden, but with your livestock as well. Summer is the time when everyone is eating well and growing quickly.

Your goats and sheep have given birth, and you may be milking them now. Your chickens are laying like crazy and you may have had a hen go broody and try to hatch out some chicks. Your bees are working full force and you're adding more supers to encourage honey production.

In the garden, your berry bushes are starting to ripen and all your warm-season plants are taking off. Tomatoes, green beans, and zucchini beckon you on a daily basis as the more you pick, the more you have to pick the next day! It's hot, though, so you're being careful to water wisely and lay down extra layers of compost and mulch as needed.

These tomatoes will produce all summer long and up to the first frost.

Stuffed Tomatoes

Garden-fresh tomatoes make this dish to die for.

Yield:	Prep time:	Cook time:	Serving size:
4 tomatoes	15 minutes	25 minutes	1 tomato

4 slicing tomatoes (such as Big Boy)

4 slices bacon, cooked, drained, and chopped

1 medium bell pepper, diced

4 TB. Parmesan cheese, grated

½ cup plain breadcrumbs

Freshly ground black pepper

2 TB. olive oil

1. Preheat the oven to 350°F.

2. Slice tops off tomatoes and scoop out seeds, which may be discarded. Use a knife to carefully hollow out each tomato.

3. Coarsely chop about ⅓ of the tomato pulp (not the seeds) and place into a bowl. Add bacon, bell pepper, Parmesan cheese, breadcrumbs, and black pepper. Mix well.

4. Fill each tomato with the stuffing mixture. Place in a small baking dish, brush or drizzle with olive oil, and bake for 25 minutes.

5. Serve immediately.

(Recipe courtesy of Home-Ec101.com)

OVER THE GARDEN FENCE

You can stuff the tomatoes ahead of time and store them in the refrigerator until you're ready to bake them. In this case, increase the baking time to 40 to 50 minutes.

In addition to the amazing tomatoes, the stars of the summer garden in my estimation, there are many warm-season herbs that are ready for harvesting. Thyme, dill, cilantro, sage, and so many others will all be harvested through the summer. Basil is one that you'll want to keep your eye on because after it goes to seed, the leaves won't be as tasty. Most herbs can be dried or frozen to preserve, but one of my favorite ways to preserve basil is in pesto.

Basil Pesto

Preserve the fleeting taste of summer while you can so you can enjoy it the rest of the year.

Yield:	Prep time:	Serving size:
2 cups	30 minutes	¼ cup

4 cups fresh basil leaves

1 cup olive oil

3 cloves garlic, minced or grated

¼ cup Parmesan cheese, grated

Salt

1. Use a blender or food processor to blend basil and olive oil. Scrape everything off the sides of the bowl and back into the middle.

2. Add garlic, Parmesan cheese, and salt to taste and blend into a pastelike consistency.

OVER THE GARDEN FENCE

Pesto freezes really well. Try freezing it in ice cube trays for quick access to small amounts. Pesto makes a fabulous pasta salad seasoning. Mix ¼ cup of pesto sauce with 4 cups of cooked pasta noodles for a delicious, summery side dish.

The other plants that really own the summer season are berries. Everything from strawberries and mulberries in early summer, to blueberries, raspberries, and blackberries will be ready to harvest through the summer season. Get your canning equipment ready and bring on the jam!

Blueberry Zucchini Bread

When you've got zucchini coming out of your ears, bake some into your bread. The sweet blueberries and moist zucchini combine to make this a mouthwatering treat you can't pass up!

Yield:	Prep time:	Cook time:	Serving size:
1 loaf	30 minutes	55 to 65 minutes	1 1-inch slice

2 eggs

½ cup vegetable oil

2 tsp. vanilla extract

½ cup granulated sugar

½ cup brown sugar, lightly packed

1 cup zucchini, shredded

¾ cup all-purpose flour

¾ cup whole-wheat flour

½ tsp. kosher salt

½ tsp. baking powder

⅛ tsp. baking soda

1½ tsp. ground cinnamon

1 cup fresh blueberries

1. Preheat the oven to 350°F. Lightly grease an 8½×4½ loaf pan.

2. In a large bowl, beat together eggs, vegetable oil, vanilla extract, granulated sugar, and brown sugar. Fold in the zucchini. Beat in all-purpose flour, whole-wheat flour, kosher salt, baking powder, baking soda, and cinnamon. Gently fold in blueberries. Transfer to the prepared loaf pan.

3. Bake 55 to 65 minutes, or until a knife inserted in the center of loaf comes out clean. Cool 20 minutes in pan, then turn out onto a wire rack to cool completely.

Variation: You can substitute all-purpose flour for whole-wheat flour, and granulated sugar for brown sugar. For **Blueberry Zucchini Muffins,** proceed with this recipe but place batter into muffin cups and shorten cooking time to 25 to 35 minutes.

(Recipe courtesy of ASouthernFairytale.com)

The root crop that takes center stage in summer is the humble potato. You can't beat the flexibility of this crop. Soups, baked, roasted, mashed, fried, and everything else. It's hard to top the potato because it combines with so many different flavors and foods well.

Red Skin Potato Salad

This delicious and cool potato salad is a nice treat for the summer.

Yield:	Prep time:	Cook time:	Serving size:
6 cups	30 minutes	8 to 10 minutes	1 cup

3 lb. red potatoes, washed and cut into bite-size cubes

½ cup olive oil

½ cup red wine vinegar

2 garlic cloves, minced

2 tsp. Dijon mustard

Freshly ground black pepper

1. Put a large pot of water on the stove and bring to boil, adding red potatoes and cooking until tender (8 to 10 minutes).

2. While potatoes cook, mix olive oil, red wine vinegar, garlic, Dijon mustard, and black pepper in a small bowl.

3. Drain potatoes and pour dressing over them and pepper to taste; mix gently to combine.

4. Chill in the fridge 1 hour. Serve cold.

By the end of summer, you'll have started preserving, canning, dehydrating, and freezing a bounty of foods. And you will have eaten like a king at the same time.

Fall Bounty from the Backyard Farm

Autumn is the season of the horn of plenty. It's when the last frantic burst of preservation happens. Winter preparations are in full force while the trees and land around you begin to slow down.

Nut trees, apples, and late berries are ready for harvest. Brambles and late fruit trees should be pruned at this time. In the garden, you'll be replanting areas you've harvested with fall crops. The potatoes come out, and spinach goes in. The corn finishes harvesting, and you plant it over with beets or broccoli. The changing season reflects itself in your garden.

Deciduous trees on your property are losing their leaves, so be wise and take advantage of that free mulch. Start a new compost bin or two with all the available plant material that's coming out of your garden and the landscape around you. Add to its richness with what your livestock offers you.

We harvest potatoes by the sack full and store them for months in the root cellar. So much deliciousness from the humble potato.

With your livestock you'll be making preparations to breed your goats and sheep. Your chickens may need to be culled if you have some hens no longer laying well. You'll harvest your honey early in the fall and watch your bees carefully as cooler weather sets in. Give everything a deep cleaning before the winter sets in. Patch any drafty areas and increase the bedding for better warmth and insulation.

Sweet Potato Casserole with Apricots

This sweet and warm casserole uses the delicious bounty of the season. Nutty flavors are enhanced by the sweet flavors of sweet potatoes and maple syrup.

Yield:	Prep time:	Cook time:	Serving size:
6 cups	15 minutes	45 minutes	1 cup

2 (8-oz.) cans sweet potatoes, drained, or 1 lb. fresh, cooked sweet potatoes (cubed)

1 (16-oz.) can crushed pineapple, drained

1 cup maple syrup

1 cup pecan halves

½ cup dried apricots, sliced

½ cup brown sugar, packed

2 TB. butter, melted

2 tsp. ground cinnamon

2 tsp. pumpkin pie spice

½ tsp. salt

1. Preheat the oven to 350°F.

2. Place sweet potatoes in an ungreased, 3-quart baking dish.

3. Combine crushed pineapple, maple syrup, pecans, apricots, brown sugar, butter, cinnamon, pumpkin pie spice, and salt, and pour over potatoes.

4. Bake uncovered for 45 minutes, or until heated through and bubbly.

Summer and winter squashes are ripening now, and you'll be harvesting squashes to store, and of course many to eat as well. Pumpkins are the usual rock stars of this season and will store for many weeks in a root cellar. Other squashes that you should not ignore in your garden include butternut, spaghetti, and acorn squash.

Simple Autumn Skillet

Here is a flexible recipe for autumn vegetables cooked with sausage. This hearty and filling dish provides immense flavor at a moment's notice.

Yield:	Prep time:	Cook time:	Serving size:
4 cups	10 minutes	30 minutes	1 cup

1 lb. smoked sausage or kielbasa, sliced

2 sweet onions, sliced

1 butternut squash, peeled and cubed

½ head of green cabbage, roughly chopped

½ cup chicken or vegetable stock

Salt

Freshly ground pepper

1. In a large, heavy pot, cook smoked sausage over medium-low heat until fat begins to render. Add sweet onions and stir, cooking for 2 to 3 minutes, then add butternut squash, cabbage, chicken stock, salt, and black pepper.

2. Reduce heat to low and cover. Stir occasionally and cook until the squash is fork tender, about 15 minutes.

(Recipe courtesy of Home-Ec101.com)

What's Available in Winter?

A lot! Below the surface of the seemingly dormant backyard is a lot of potential food and crafting supplies if you know how to find them. Not only will many of your fall crops store well throughout the winter, but your fall garden can continue to produce for a long time. In some climates, through the whole season. Not to mention how you can take advantage of the food you have stored.

Applesauce Oatmeal Coffee Cake

Warm your kitchen and the hearts of your loved ones as well. Sweet applesauce gives this comfort food a delicious taste.

Yield:	Prep time:	Cook time:	Serving size:
1 cake	20 minutes	35 to 40 minutes	1 1-inch slice

3 cups quick or old-fashioned oats	2 tsp. baking powder
1½ cups unbleached flour	1 cup plus 1 TB. brown sugar
1 tsp. allspice	1 cup milk
1½ tsp. baking soda	2 eggs
1 cup whole-wheat flour	2 cups applesauce
¾ tsp. cinnamon	6 TB. vegetable oil

1. Preheat the oven to 375°F. Combine oats, unbleached flour, allspice, baking soda, whole-wheat flour, ½ teaspoon cinnamon, baking powder, and 1 cup brown sugar.

2. Mix together milk, eggs, applesauce, and vegetable oil and add to the dry ingredients, stirring just until moistened.

3. Pour into a 9×12 baking pan.

4. Combine remaining 1 tablespoon brown sugar and remaining ¼ teaspoon cinnamon. Sprinkle topping over batter.

5. Bake 35 to 40 minutes, or until a knife inserted in the center of cake comes out clean.

6. Cool for 10 minutes. Tipping the loaf pan on its side during cooling will make it easier to remove.

OVER THE GARDEN FENCE

Double or triple this recipe when you have lots of fresh applesauce. The batter will freeze for a quick breakfast another day.

Anything in the *Brassica* family, such as cabbage, kale, and kohlrabi, can be grown successfully through the fall and winter. In fact, kale tastes better after a good frost than it does beforehand. The trick with kale is to de-stem the kale leaves first. I like to slice up the leaves after that and use them for breakfast. You can sauté them and mix them in your morning eggs. Or boil them until soft and blend with homemade yogurt, honey, and maybe some berries from your freezer to make a great breakfast smoothie.

On the livestock side of things, you'll have your bees in their winter cluster. Chickens will have ceased or decreased their laying with the shorter winter days. And most of your meat stock that was from last year has probably been butchered and put in the freezer now, giving you a freezer full of tender, nutritious meat.

Roast Pork Tenderloin with Winter Greens

This can work with any roasted meat or greens, so mix it up and use those green veggies. The homegrown meat has a deep flavor not possible elsewhere, and the nutritious greens have a sweeter flavor in the cool seasons.

Yield:	Prep time:	Cook time:	Serving size:
1 1-pound tenderloin with about 1 cup greens	20 minutes	20 to 25 minutes	¼ tenderloin with ¼ cup greens

1 lb. pork tenderloin	1 medium onion, thinly sliced
⅛ tsp. plus ¼ tsp. salt	1 lb. kale or other winter greens, tough stems removed
Freshly ground black pepper	½ cup chicken stock or low-sodium chicken broth
2 cloves garlic, minced	
4 tsp. olive oil	1 TB. red wine vinegar

1. Preheat the oven to 425°F.

2. Line a baking sheet with aluminum foil. Rub the pork tenderloin with ⅛ teaspoon salt, black pepper, and garlic.

3. Heat a large, heavy skillet over medium heat. Add 2 teaspoons olive oil and pork to hot skillet and brown on all sides (3 to 4 minutes). Remove pork tenderloin from the pan and place it on the baking sheet. Place the baking sheet in the oven and cook until tenderloin reaches 145°F, about 12 minutes. Cover loosely with foil and allow pork to rest for 3 to 5 minutes.

4. Add remaining 2 teaspoons olive oil and sliced onion to hot skillet. Cover and cook until onions soften and just start to brown. Add kale, chicken stock, remaining ¼ teaspoon salt, and additional black pepper to taste. Cover and cook, stirring occasionally, until kale is tender, about 5 minutes. If after 5 minutes there is a lot of liquid in the pan, cook uncovered until most of the liquid has evaporated.

5. Stir in red wine vinegar and remove from heat.

6. Slice pork tenderloin into medallions and serve with kale and onions.

(Recipe courtesy of Home-Ec101.com)

Butchering on the Backyard Farm

The truth for backyard farmers who aren't vegetarian is that our livestock provides for us and our families as much as we provide for our livestock. And for our family, we're committed to eating humanely raised and butchered meat as much as possible. Which means that most of what we eat, we've raised ourselves. This gives us the control to know what, if any, hormones and medications are being used on the animals we eat, what quality of food they have, what quality of life they have, and the manner in which they are butchered at the end.

It's about choosing to put aside my temporary discomfort in that part of the process, to make better choices for my family, and ultimately, the animals themselves. I know that some people think it's cruel but most Americans eat meat. My family eats chicken. I would rather that chicken be one that I know had a happy, very healthy life than one that I know nothing about because it grew up in an overcrowded, overmedicated, and stressed environment and was treated with disdain and often cruelty its entire life.

Overview of Processing Meat and Chickens

Some of the meat processing we do ourselves. Chickens in particular are very easy to self-process. And other animals, like the male goats that are born to us each year, we often send to a local butcher for processing.

Butchering involves killing the animal, skinning or plucking the feathers, and preparing the meat to be able to be easily cooked and eaten. I was very squeamish the first time we processed a handful of chickens but now, while it makes me sad to have them killed, I am appreciative of the ability to feed my family in a positive and eco-friendly way.

There are a couple different ways to dispatch the chickens. One way is to use a cone which is like a large funnel. The chicken goes head-first into the cone and its head sticks out of the bottom opening, allowing the neck to be cut quickly. The other way is the old-fashioned way of chopping off the head with a well-placed axe blow. We've found it easier to use the axe as it stresses out the chickens to be hung upside down in the cone and takes longer to get them ready. If you aren't handy with an axe, though, I don't recommend using this method. (I should note here that I make my husband or father-in-law do this part and I take over after the deed is done.)

After that part is done, we plunge the bird into boiling water, which helps to loosen the feathers. And then we pluck all the feathers off. Some people will just skin the chicken completely rather than pluck the feathers. After that, the insides are carefully removed and discarded. If you want to quarter the bird into smaller pieces you can, although roasters can be kept whole. Either way, rinse the meat clean, pat it dry, and get it in the freezer as quickly as possible.

Plucking feathers is a messy job, but easier done with the help of boiling water. It only takes a few minutes and provides the highest-quality meat possible for your family.

The process for rabbits, goats, and sheep is similar. The animal has to be dispatched, skinned, cleaned, and then the meat cut into pieces appropriate to the animal. After that, the meat is cooled and placed in long-term storage.

Finding Local Help

Many beginners aren't sure how to start processing their own meat, or just flat don't want to do it. There are other options available but they all involve a little more (sometimes a lot more) money. You can take the animal to a local butcher or meat processor. We have a local source that we really like because we are always sure to get our own meat back (see the following sidebar).

Expect that the cost of having someone else process your chicken for you may double or even triple the per-pound price of the meat. The chickens are so small compared to something like a goat or hog, that even if a processor only charges $5 per bird that could triple the cost of the meat you end up with in the freezer. The same would apply to rabbits. These smaller animals may be a lot more cost-effective to process yourself.

Larger animals obviously need more room for processing than most people would have access to in their backyard. And that's where a local meat processor will come in handy. Check around the feed store or local hunting supply stores for local butcher recommendations.

THORNY MATTERS

One year we used a larger butcher house to process a deer and we ended up getting back mixed meat—that is, we got some meat that belonged to a deer someone else had killed. I don't like the idea of investing in our livestock specifically to feed our family, only to get meat back that I don't know anything about! So wherever you take your animals to be processed, please find out what procedures they have in place to ensure that your livestock is the meat that's returned to you. A reputable, USDA-inspected processor will be happy to explain their system to you.

There's an incredible gap in availability in some areas that do not have a local butcher nearby. In this case, I would recommend trying to find someone who has experience with processing their own animals. Check your local Community-Supported Agriculture (CSA) facility or farmer's markets for anyone selling locally raised meat and find out where they got their processing done. If they are processing the meat themselves you might see if they'd be willing to let you partner up with them on their processing days.

One interesting trend beginning to develop is the mobile slaughterhouse. A large trailer serves as the processing plant and the local butcher can literally come to you! There are only a dozen or so operations at this time, but I have a feeling the trend will catch on as more people seek homegrown foods for health and financial reasons.

Learning to eat what is available, when it's available, means you're eating the freshest, most nutritious food you can possibly find. Whether fruit, meat, or vegetables, a bountiful feast can be found right out your back door.

Canning and Freezing Produce and Meat  17

One of the things that happens when you grow your own food is that you'll likely have an abundance of one produce at any given time. Your tomatoes will all be harvested over a three-, maybe four-month period, yet you want to be able to produce in those few weeks all the tomatoes you will use for the entire year. Storing the tomatoes so you can use them during the winter months when nothing is growing makes your food supply a year-long provision for your family.

Canning and freezing are two of the primary methods of storing excess produce from the backyard farm and are two of the simplest methods for beginners. While most people tend to think of canning vegetables, I find the taste much better when veggies are preserved by freezing instead. Canned vegetables tend to be good only for stew and chili, where their soggy texture won't make much difference.

Fruit, on the other hand, is wonderful frozen, and blanching vegetables before freezing means you wouldn't have to cook them as long. Most foods will store in the freezer 6 to 12 months, while canned goods will store, without electricity I might add, for two or three years on the short side.

Try experimenting with various types of food preservation for the different types of food you want to save and see what works best for you. Having long-term storage in place is a key component of homestead life. Food is cyclical on the homestead so you grow what can be grown when it can be grown and store as much as possible for when it isn't available fresh. This mind-set was one of the biggest adjustments for me when I began my journey to self-sufficiency. Not depending on the grocery store for shopping on a weekly basis was a foreign concept that took some getting used to. Becoming more self-sufficient means storing the bounty of the backyard farm at the peak of the seasons, when bounty is available.

When you harvest this much, or more, every day, you need a way to preserve it. Canning is still one of the best ways to preserve extra food.

Equipment Needed for Home Canning

It's important to have the proper equipment on hand when you are canning your meat and produce. Canning is very safe, but only when done properly. The exact equipment needed depends on the type of canning you plan to do: *water-bath canning* or *pressure canning*. I like to have equipment on hand for both canning techniques. (I discuss the steps involved in both methods later in this chapter.)

DEFINITION

Water-bath canning is accomplished by immersing the cans in boiling water. The least-expensive method of canning, it is only useful for high-acid foods such as fruits and pickles. **Pressure canning** uses a pressure canner to increase the temperature during canning and is used for low-acid foods like most vegetables and meat.

Before you begin canning you should have all your equipment at hand so you don't have to stop halfway through the process. You can reuse your mason jars and rings, but should inspect your mason jars carefully for any sign of chips or cracks that could cause problems sealing. You'll need fresh lids each time, but they are a minimal investment once you have the jars and rings.

A pressure canner is on the left, and a water-bath canning pot is on the right.
(Photo courtesy of Daniel Gasteiger)

These are the items you'll want to have for canning:

- Glass mason jars in various sizes

- Metal screw bands or "rings" to hold the lids in place

- Metal lids with sealing compound

- Deep pot for water-bath canning

- Pressure canner (see the next section for more about the different types)

- Jar rack to suspend cans in boiling water, and lift them out

- Jar lifter to lift individual jars

- Canning funnel for filling jars (a nice-to-have item)

- Lid-lifter magnetic stick to lift dropped lids and rings out of the water (a nice-to-have item)

- Lid wrench, also called a band tightener, if your grip is too weak to tighten manually

- Rubber spatula or other bubble releaser to get rid of trapped air in the prepared food (many of these will have marks to measure headspace)

- Miscellaneous hot pads, dish rags, and towels for cleaning and moving the jars

Other equipment and kitchen utensils might be used when canning specific items. A saucepan for heating lids and rings, for example. Also a strain cloth for jelly might be needed. I'll discuss specialty items as I discuss what they would be used for.

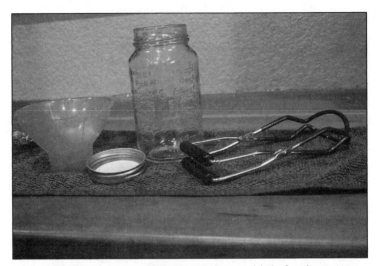

Have all your canning supplies clean and accessible before beginning.
(Photo courtesy of Brannan Sirratt)

Pressure Canners

Pressure canners allow food to be cooked at a higher temperature than water-bath canning by creating an airtight seal and preventing the steam from escaping. If water at sea level boils at 212°F, a pressure canner at 10 pounds of pressure will not boil until 240°F.

There are two main types of pressure canners: weighted-gauge canners and dial-gauge canners. Weighted-gauge canners are the most simplistic in design and operate by placing a weighted disk over the steam vent covering. In the pressure canners with adjustable weights, these vent covers may come in 5-, 10-, and 15-pound sizes. The vent cover will raise and allow steam to vent after the pressure has built up to a high-enough level to displace it. The weighted-gauge pressure canners are the ones that have the pieces rattling back and forth on the lid—that's the vent cover.

Dial-gauge pressure canners are more precise in the amount of pressure because the needle points to the exact amount of pressure. This allows you to control the exact amount directly. The vent pipe does have a weighted safety release; however, that is usually set at 15 pounds. This acts as a back-up safety in case you aren't able to control the pressure canner or the dial malfunctions. If you ever have doubts about the accuracy of your dial gauge you can have them tested at your local extension office for a small fee.

A weighted gauge on the lid is used to control pressure in some pressure canners.
(Photo courtesy of Daniel Gasteiger)

Regardless of the style of pressure canner you purchase, there are some elements they all have in common:

✢ The lids of all pressure canners have airtight seals and usually come with a *rubber ring* or *gasket.* These need to be replaced if there are any signs of damage that would prevent a good lock between lid and cooker.

✢ All pressure canners have an *air vent lock* (a.k.a. exhaust) that allows the steam to escape when you are done cooking. Always check to make sure the air vent is clear and not stuck closed so the pressure can be relieved as needed. Check your manual to find how long you need to vent the steam before it is safe to remove the lid. You don't want to take the lid off and be scalded immediately by the heated steam.

✢ Your pressure canner should also have an *overpressure plug* that will act as a fail-safe to prevent accidents with too much pressure build-up.

For home canning you probably want a pressure canner that will hold at least 7 quart-sized jars at a time. There are models that hold 9 quart-sized jars and even some that hold 14 quart-sized jars, but the larger sizes are less flexible and take longer to heat. For most backyard farms the 7- or 9-quart models are sufficient.

OVER THE GARDEN FENCE

You might decide to try your hand with canning using a water bath and then graduate to pressure canning. A pressure canner can be useful year-round, not just when you are canning in the summer and fall. A high-quality pressure canner will last for years and can also be used as a pressure cooker for cooking food quickly while keeping it tender.

Home Canning Safety Tips

The reason for using a pressure canner to can many vegetables and other produce is to prevent food poisoning from botulism. There is a bacteria often found in the soil and thus on many fruits and vegetables called *Clostridium botulinum.* Usually it is harmless and won't affect you, but when it gets wet the bacteria becomes a harmful, infectious agent that causes food poisoning known as botulism. That's why when food is canned the bacteria has to be killed. Either the food has to have a high-enough acidity level to kill the botulism spores (high-acid foods), or else the temperature has to reach a level high enough to kill the spores, which is where the pressure canner's 240°F comes in. It is important that the lower-acid foods you can are cooked at 240°F for long enough to thoroughly kill the bacteria. And once the bacteria is killed, the cans are only safe long-term if they seal properly and aren't opened.

Pressure canning is really quite easy to do, but as with any kitchen activities, care should be taken:

⚜ The high temperatures mean you should be careful not to get burned. The pot, the lid, the steam, and the jars get very hot. Make sure you fully vent the heat before removing the lid.

⚜ Clean any glass jars before using them. Even new jars should be cleaned before use. Double-check the rims of the jars to make sure there are no chips or cracks because those weaknesses could cause the jar to break, or just not seal. Never reuse jars that you bought food in from the grocery store (such as mayonnaise jars) as they usually aren't suitable for home canning. You can reuse the jars that you get for home canning, though, until they crack or develop chips.

⚜ Sterilize the lids and rings before using them by boiling in hot water for 10 minutes. I usually boil the jars to clean them as well. Leave the rings and bands in the hot water after you turn off the boil so they are still warm when you apply them.

Note that some vegetable preparation methods do not yet have a USDA-approved method of canning. Mashed squash, pumpkins, potatoes, and sweet potatoes should not be canned. Use other preparation techniques like cubing for these foods so you can be sure of your cooking times. If you want to mash them later, you can used the canned, cubed pieces and blend them to mash them right before using.

THORNY MATTERS

When you pull a can of produce from your larder, always inspect it carefully for signs of damage or contamination. Is the lid still tightly sealed? If not, throw it out. If there are signs of foamy bubbles in the can, moldy spots present in the food, or a bad smell to the food, you should discard the food immediately without tasting it.

Basic Steps for Water-Bath Canning

Water-bath canning is a method that is only safe for high-acid foods like fruit, tomatoes, and foods in vinegar like pickles. Canning fruits at the moment of harvest is one of the best ways to preserve the maximum level of nutrients. Be sure you find a recipe to follow based on USDA safety recommendations.

Here are the basic steps for water-bath canning:

1. Preparing the food for canning means peeling, pitting, deseeding, or in some cases precooking the fruit. Discard any soft, rotten, or spoiled items. Some foods are put into the cans unheated, known as *raw* or *cold packed,* while other foods are put into hot jars precooked and heated, known as *hot packed.* Each recipe specifies the method of preparation, how much headspace (empty room) to leave at the top of the jar, and how long to process the jars in the pressure canner.

2. Fill the canning pot with water until the level is within an inch of the top of the jars. Bring the water to boil to heat the jars. You should also be warming up the lids and rings at this time but not to a full boil.

3. Fill the hot canning jars with the produce—either hot sauces or raw fruit chunks, and boiling liquid, as directed by the recipe. Fill the jar to the appropriate level, leaving the correct amount of headspace at the top according to the recipe. Use your bubble releaser or a rubber spatula to make sure there are no air bubbles trapped around the fruit in the jar.

4. Wipe the rim of the jar clean with a clean dish rag or towel and place a hot lid over the top of the jar. Screw on a ring and hand tighten.

5. Place each jar into the water-bath canner carefully. Use a jar lifter to grasp the glass of the jar and avoid messing up the band. When all the filled and lidded jars are in the canning pot, on the canning rack, make sure there is at least an inch of water over the top of the jar.

6. Put the lid on the water-bath canner and bring the water to a boil. Remember that when canning recipes you start the timing at the point of boiling so if the recipe says three minutes, that means *boiling* for three minutes.

Be careful when taking the lid off the water-bath pot—it will be hot!
(Photo courtesy of Brannan Sirratt)

7. Carefully remove the jars when the time is up and take care to keep them upright and not jostle them about. Set the jars on a towel or cooling rack and make sure none of the jars are touching.

Use a jar lifter to carefully remove the jar so you don't burn yourself.
(Photo courtesy of Brannan Sirratt)

8. Within 10 to 20 minutes the jars should be sealed. The lids will pop downward in the middle, signifying a good seal has been achieved. Let the jars fully cool before putting them away in the pantry after testing the seal to make sure it's secure. It's a good rule of thumb to leave them sitting overnight to ensure a good seal.

If any of the jars fail to seal, refrigerate them and use the contents within a week. Be sure you mark or label the jars so you know what you've canned. Trust me, it's not always easy to tell and you'll want a way to include the date so you know when the produce was canned.

Basic Steps for Pressure Canning

Pressure canning is the method used when the food being prepared is not acidic enough to kill the botulism. It allows the temperature to be raised high enough to kill potential mold and bacteria as well. Pressure canning techniques are very similar to the water-bath canning techniques so you'll notice a similarity in the steps. Before starting, check your equipment, jars, and pressure canner seals as described earlier in the section "Home Canning Safety Tips" and in your user manual.

For more information and specific instructions on canning and other methods of food preservation, a good resource is *The Home Preserving Bible (A Living Free Guide)* by Carole Cancler.

Canned carrots with dill are made in the pressure canner because carrots are a low-acid food.
(Photo courtesy of Broadfork Farm)

Here are the basic steps for pressure canning:

1. Heat your lids and rings in a pot of hot, but not boiling water. Fill your jars with hot water in preparation for packing.

2. Prepare the produce or meat that you'll be canning according to the recipe provided (whether raw packed or hot packed) and begin heating the water in the pressure canner so it's warming up and ready to begin boiling when the jars are filled.

3. Pack the jars with your produce for canning and use your bubble releaser or rubber spatula to remove any bubbles from the mixture. Check that you have the right amount of headspace at the top of the jar as specified by the recipe.

4. Put a lid and ring on each jar and hand tighten. Your canning pot should be on the stove with boiling water in it now. Add your canning rack and place the full jars on the rack. You'll probably only need about a quart of water in the bottom of the pressure canner, but check the owner's manual that came with your canner to verify the amount.

These jars of corn are full and ready for processing in the pressure canner.
(Photo courtesy of Daniel Gasteiger)

5. Secure the pressure canner lid and place over high heat. Turn the burner up until the steam begins to blow through the pot's vent pipe. Watch the level of the pressure build up until it reaches the pressure given in the recipe. Remember that timing begins when the pressure is at the correct level.

6. The pressure must be maintained during the entire canning process. If the pressure falls below the target level you have to restart your timing from the beginning.

The pressure canner is working—here the dial gauge shows just over 10 pounds of pressure.
(Photo courtesy of Daniel Gasteiger)

7. Let the pressure canner cool completely before removing the lid. This can take 30 to 45 minutes at least. Use caution when removing the lid.

8. Remove your finished jars from the pressure canner, keeping them upright, and place them on a cooling rack. Be sure none of the jars are touching so there is space between them to thoroughly cool.

9. Check your lids and make sure they have properly sealed. Don't push on the tops with your fingers, or you may create a false seal. You'll hear a "pop" and should notice the middle of the lids pushed down, in toward the middle of a jar. Refrigerate any jars that don't seal and use the contents within a week.

Don't forget to label your jars with the contents and date! It's easy to forget when exactly you canned something and you want to be able rotate through your stock.

GETTING HELP

If you are nervous about getting started, find someone with experience to partner with. I bribed my mother-in-law to teach me how to can by bringing over all the supplies and produce, and giving her half of the canned goods from our first session. More and more communities are forming preparation groups. Instead of each homesteader (or homesteader wannabe) trying to do all the canning themselves they can combine forces with other people. Like an old-fashioned barn-raising, the group effort saves energy, keeps everyone entertained, and gives newbies a chance to learn from others who are more experienced.

Freezing Tips and Tricks

Often, a large part of putting food away when you have it available, to use at a later time, is having an extra freezer on hand. While it may seem like an expensive investment, especially with the energy costs of maintaining the freezer, the numbers look pretty favorable! We bought our chest freezer for about $400 and it will last 15 years. A monthly energy cost is projected to be less than $10 per month. This means our yearly cost is less than $140 for the cost of the machine and the energy costs combined. It would be almost impossible not to save $140 per year on your grocery bills using a deep freeze.

Chest-style freezers are more energy efficient than the stand-up units because the cold air does not pour out of the freezer every time you open the door. It is also cheaper to maintain a freezer that is full than it is to maintain a freezer that is empty. So if you have it, use it!

We have a gas generator as a backup for our freezer in case of a power outage. If your power does go out, do not open the freezer door. Your food will stay cold for up to two days. After that you risk losing some of the frozen foods, especially the foods near the top and edges of the freezer.

When we are low on meat or vegetables to freeze, I will sometimes prepare large batches of meals and freeze them precooked. We also buy staples like wheat for grinding into fresh flour, and rice, in bulk for the deeper discounts, and storing some of this in the freezer when there is room allows us to keep them bug free and keep the freezer full at the same time.

Freezing vegetables instead of canning them keeps them more fresh-tasting than canned vegetables. But of course, the trade-off is that frozen vegetables don't last as long as canned vegetables do. Many crops have to be partially cooked in boiling water, a method called *blanching,* to slow down the enzymatic processes of the produce.

Freezing can change the texture of the fruits and vegetables you freeze, as well as causing some discoloration. You can coat fruit with a mixture of ascorbic acid and water, or water mixed with salt, before freezing to help prevent browning. Freezing items as quickly as possible can also help minimize the cellular changes that happen when the items freeze.

When you freeze food, I've found it best to prepare the fruits or vegetables first: peeling them, chopping them, removing stems, leaves, or pits, etc. You can freeze food items in single-use portions so they will be close at hand when cooking.

Alternatively, you can freeze the vegetable items on a flat cookie sheet and when they are frozen, put them into a larger container. By doing that your frozen produce won't stick together in one big clump. This works especially well with berries.

If you are freezing liquids like sauces or soup stock, you can freeze them in gallon-size, zipper-lock storage bags placed inside plastic containers. This will allow the liquid to freeze in a flat, easily stackable shape. When the liquid is frozen, you can remove the containers and stack the bags in the freezer. This technique will save you a lot of room and allow you to maximize every cubic foot of freezer space.

FREEZING BERRIES

To freeze berries, wash them thoroughly and pick out any stems or leaves. Gently pat the berries dry and lay on a flat pan or cookie sheet. Place them in the freezer until each berry is individually frozen solid. When each berry is frozen singly, you can place them together in a container and they won't stick together. Freezing berries from your backyard allows you to preserve berries that are not otherwise available year-round. Tender berries such as raspberries and blueberries are not able to be shipped long distances. Being able to handle each berry by hand, with a tender touch, makes all the difference.

Vegetable Blanching and Freezer Preparation Chart

Blanching vegetables before freezing slows the ripening process and will help preserve as many of the nutrients as possible. Start with fruits and vegetables at their freshest moment. Bring the water to a boil before adding the vegetables.

Steam or boil the produce according to the suggested times (see the following table) and then immediately remove. This is easier if you use a strainer in the boiling water. Cool the vegetables immediately—I usually have a sink full of ice water ready and plunge the produce in it to cool as rapidly as possible.

When the produce is cooled, drain it thoroughly. Then you can freeze it in a bag or freezer container, or individual units can be frozen loose on a tray and added to a large bag later. Don't forget to label your freezer bags and containers! Include what the item is, how it's prepared (chopped, whole, grated, etc.), and when it was frozen.

Cooking frozen foods doesn't require as much cooking time as using raw foods. Some vegetables like corn on the cob and all fruits benefit from being partially defrosted before cooking.

Vegetable	Blanch or Cook Time
Asparagus	Blanch 2 to 4 minutes depending on size
Beans (green, snap)	Blanch 3 minutes
Beans (lima)	Blanch 3 to 4 minutes
Beets	Cook completely
Broccoli	Blanch 3 minutes
Cabbage	Wedges, blanch 3 minutes; shredded, blanch 1½ minutes

continues

continued

Vegetable	Blanch or Cook Time
Carrots	Blanch 2 minutes
Collards	Blanch 3 minutes
Corn (sweet)	Whole kernel, cook 4 minutes; on cob, cook 7 to 11 minutes
Eggplant	Cook completely, then freeze
Greens, noncollard	Blanch 2 minutes
Peas	Blanch 1½ to 3 minutes
Peppers, bell	Blanch 2 to 3 minutes
Potato	Cut into ½-inch cubes and blanch 5 minutes
Pumpkin (winter squash)	Cook completely, then freeze
Summer squash	Slice ½ inch thick and blanch 3 minutes
Tomato	Cook completely, then freeze

Packaging Meat to Freeze

One of the best uses of the freezer for us has been freezing large portions of meat. When you butcher a couple dozen chickens or a young male goat culled from the dairy herd, you end up with 100 pounds of meat at one time. Using your freezer allows you to save these resources for using later.

Even if you don't butcher your own livestock, you can purchase meat in bulk portions and save a significant amount of money. You'll be able to purchase locally raised meats and get the freshest food for your family.

Use the freshest meat available and before you freeze it, process it into the cuts and pieces in sizes you'll use. You don't want to have to thaw an entire front shoulder just to get a small roast. Have plenty of freezer paper and tape on hand for processing the meat, and keep a permanent marker close by to label the package with the cut and date.

THORNY MATTERS

Clean your hands and the preparation area thoroughly before wrapping your meat for the freezer. You don't want to introduce contaminants to your meat. Of course, you should always thoroughly sanitize an area where raw meat has been handled after you're finished.

A butcher wrap technique provides a good covering for the meat to help prevent freezer burn. Place a good-size sheet of freezer paper on a flat surface like a counter or table. I use an 18- to 24-inch-long piece to wrap a pound of ground meat.

Tuck the first corner against the meat and smooth it down so it makes good contact.

Place the ball of meat near one of the corners and tuck the corner up against the meat. Next, fold the side corner up and hold them in place with your hand. Press the paper down against the meat to help eliminate any air.

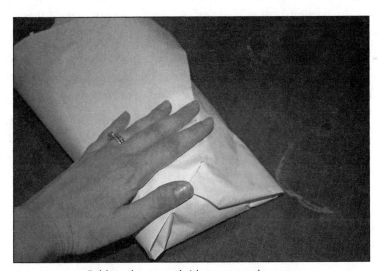

Fold up the second side corner and secure.

Bring the sides in, and then roll the meat over, allowing the paper to cover all sides of the meat.

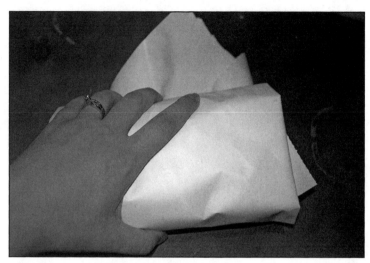

Roll the meat over and wrap securely.

Use the freezer tape to secure the paper and label the package with the cut and date. Ta-da! Your finished package is ready for the freezer.

If you minimize the air contact to your meat, and use paper or bags specifically designed for long-term freezer storage, most meats will hold their flavor for a full year. Beyond that it isn't an issue of safety, but rather of flavor, in using the meat. The wrapping job makes a big difference, so be sure you remove as much air as possible.

Preserving food through freezing and canning can help you save money. Putting your food away while there is a bounty keeps you well fed and happy during the lean times.

Freezing premade meals is another way to freeze meats and vegetables. When cooking the bounty of the backyard farm, try cooking larger portions or doubling the meals you make. For example, when I make the following Inside-Out Lasagna recipe, I make three or four pans instead of one pan. One pan is eaten that night, and the other pans are stored in the freezer for quick-fix meals.

Erica's Inside-Out Lasagna

This cheesy pasta dish that also incorporates the full flavors of homegrown meats and summer-ripe tomatoes can't be beat!

Yield:	Prep time:	Cook time:	Serving size:
16 rolls	30 minutes	15 minutes	2 rolls

1 (16-oz.) box lasagna noodles

3 cups chopped vegetables of your choice (I use celery, onions, and carrots)

1 lb. ground beef

8 oz. spaghetti sauce (store-bought or homemade)

8 oz. alfredo sauce (such as Ragu Cheesy Classic Alfredo Sauce)

8 oz. shredded mozzarella cheese

Fresh parsley sprigs for garnish (optional)

1. Preheat the oven to 350°.

2. In a large pot of boiling water, cook lasagna noodles until almost done, but still firm and workable. Cut each noodle in half. In a large skillet, sauté vegetables until tender, adding beef halfway through the process. Cook until meat is browned and drain fat.

3. Add spaghetti sauce to meat and veggie mixture until it sticks together nicely. You don't want this part too soupy.

4. Pour 4 ounces alfredo sauce in the bottom of a large baking dish.

5. Place half a lasagna noodle on a plate and spoon a few tablespoons of meat, veggie, and sauce mixture in the center. Gently roll up and place in a large baking dish.

6. Continue this process until the baking dish is full. Top with remaining 4 ounces alfredo sauce and shredded mozzarella cheese. Garnish with parsley (if using).

7. Freeze in the pan, or bake until cheese on top is melted and dish is hot all the way through.

(Recipe courtesy of UntrainedHousewife.com)

Dehydrating and Smoking ∽18

One of the oldest methods of preserving food was to smoke and dry the plants and meat for long-term storage. Before electricity, humans were sun-drying herbs while they were in season, so they could be used later in the year. Smoking was another method of preserving meat and is still used today to enhance flavor.

These forms of food storage and preparation all involve removing moisture from the foods—whether it is meat, herbs, vegetables, or fruit. Historically this involved smoking meat in smokehouses, laying herbs in the sun to dry, and dehydrating fruit like grapes to be stored as raisins.

Technology makes it easier now. We also have a resurgence of some traditional methods like the solar oven, which allows us to harvest the passive energy of the sun instead of using electricity. Dehy-drated foods such as dried herbs and beef jerky only need to be stored out of the air and moisture, which means any shelf out of the sun will do. Having food that doesn't rely on refrigeration can be a benefit in emergency situations.

Drying Herbs

Probably the dehydrated food that people are most familiar with is dried plant matter—most of the spices in your kitchen are herbs that have been dried and chopped into small pieces or ground into powder. But spices are pricy, and often by the time they reach your kitchen pantry they have lost much of their flavor and nutrients.

The simplest way to dry herbs is to simply hang them upside down in a cool, dry place. If the herb is an annual, sown anew each year, you can harvest the entire plant and hang it upside down. If you're harvesting perennials, just prune the stems you need and hang them in bunches. The ideal place to hang your herbs is somewhere with good airflow but out of any direct sunlight.

Dehydrating fresh herbs, like this parsley, helps preserve them for future use.
(Photo courtesy of Brannan Sirratt)

Tie the stems of a bundle of herbs together and hang them to dry.
(Photo courtesy of Brannan Sirratt)

By removing the water, you are removing the bulk of the weight and space in the plant. Storing now becomes much easier as the herbs take up much less space.

OVER THE GARDEN FENCE

If you store herbs in small pieces, they will retain more flavor but take up more space. Grinding the herbs into a powder is the most efficient use of space, but the flavor won't last as long.

Choosing a Dehydrator

Buying a dehydrator will probably be an investment for your backyard farm that pays for itself in the first year or two. Depending on the size of your homestead, you'll want to make sure your dehydrator is large enough to handle an influx of produce at the peak of your season.

The small circular models are inexpensive (I've seen them for less than $50) and might be a good choice to test these methods. However, these inexpensive models may not always hold up under long-term use and can dry unevenly. The more sturdy models are made of higher-quality plastic or cool-touch metal frames.

Some people prefer square dehydrators with shelves that can be pulled out individually and cleaned. The benefits of this style are that shelves can be removed if you have bulkier items that wouldn't fit in the stackable shelves. The shelves are also easy to clean and replace. Stackable dehydrator units don't have single shelves that fit into a frame, but rather the shelves themselves fit together and create the entire structure. These higher-end models can run as much as $400 or more, so it might be best to start with a cheaper model to see what features you really need.

Whichever model you choose, you'll want to make sure there are fans in place to blow the hot air throughout the dehydrator. Some box dehydrators have heating elements and fans in the back of the unit so the heated air flows evenly over each tray. This helps speed up the overall drying time and also ensures that all the trays process at the same speed. The higher the wattage of your dehydrator, the faster you'll be able to dehydrate your food. Some dehydrators don't have fans to blow the hot air around but rely on convection heat instead.

Stackable rings can dry unevenly because the heating elements are usually at the bottom of the unit. So you may have to switch out the rings to allow each ring to dry in the same time frame.

For use on the backyard farm it is nice to have a unit with a wide range of adjustable temperature settings. A timer can also be beneficial but isn't a must-have. You also want to make sure that the model you pick has all the accessories you need. You'll surely want the ability to make fruit leathers, dry plants with small pieces (you'd need a fine mesh screen for this instead of a regular screen), and create jerky.

Dehydrating Fruits and Vegetables

Fruits and vegetables can be dehydrated when in season and enjoyed through the rest of the months when it isn't being harvested. Here are some basic steps for dehydrating fruit and vegetables in a dehydrator. You'll want to check your dehydrator's manual for specifics:

1. Prepare the food you are going to dehydrate. For most fruits this simply means peeling and slicing them ½ inch thick unless the fruits are small. Vegetables need to be sliced and then blanched. See the suggested blanching times for vegetables in the table in Chapter 17. Your dehydrator should also be prepared as well, which means cleaning it and making sure it's been sanitized since the last use.

2. Fill the trays with the produce you want to dehydrate and set your dehydrator at the proper temperature for the produce you are dehydrating. The temperature will be much lower for fragile herbs, about 90°F. The setting will be much higher for thicker fruits and vegetables and likely fall in the 140°F range. Don't mix freshly added plants in the same tray as partially dehydrated plants—try to dehydrate full batches at a time.

3. Check on the progress of your drying plants at the earliest time of its range. For example, if the manual calls for 6 to 8 hours drying time, you'll want to check the fruit at 6 hours time and then every 30 to 45 minutes after that. Herbs and leafy plants should be crisp, with no moisture left in them. They should crumble when you squeeze them. Vegetables should also be completely dried and brittle to the touch. Fruit slices, on the other hand, are better off slightly chewy, like a raisin would be, instead of brittle (the natural fruits will act as a preservative).

4. Package your dried produce in airtight containers such as jars, canisters with rubber seals, or zipper-lock storage bags. When it comes to stored food, the enemies are light, moisture, and heat. Room temperature is fine for dehydrated foods, which is one of the benefits of preserving them with this method.

MAKE YOUR OWN VEGETABLE BOUILLON

Use the less-than-perfect vegetables from your home garden, items you cannot sell or don't want to add to your own salads, in the dehydrator. When the vegetables are dried, they can be ground into powder and mixed together to create a homemade vegetable bouillon mix. Just add water to the powder when you're ready to use it! Try vegetables such as celery, asparagus, onion, tomato, carrots, peppers, or whatever else you happen to have on hand. Herbs like thyme, oregano, basil, and parsley are great additions as well. Add salt to the mix at a ratio of ½ veggie mix and ½ salt (we use sea salt), and you've got your own vegetable stock powder you can use in soups and other dishes.

Some of the most expensive snacks you purchase from the grocery store can be made at a fraction of the cost at home—and more nutritiously! The following fruit leather recipe is perfect for preserving short-lived berries, excess apples, or delicious summer plums and pears. It will replace fruit rolls you buy in the grocery store that are laden with artificial flavors and food colorings.

Fruit Leather

Feel free to change this basic recipe any way you like. Apples make a great base for fruit leather because they stay flexible when dried. Mix with other fruit, such as plums and raspberries, to change the flavor, and experiment with the amount of sugar and spices according to the fruit you use and your tastes.

Fresh fruit (about 8 cups chopped)	Sugar (amount will vary)
1 cup water	Spices of your choice, such as
2 tsp. lemon juice	cinnamon, nutmeg, or allspice (optional)

1. Remove stems, peels, and pits or seeds from fruit. Chop fruit flesh into small pieces. Place fruit pieces into a large saucepan and add water. Bring to a slow simmer.

2. Let fruit simmer for 10 to 15 minutes until soft, and mash with a potato masher. Add lemon juice to fruit; this ensures the color doesn't change during dehydration. Taste fruit mixture and add sugar and spices (if using) to taste.

3. Allow fruit mixture to thicken for another 5 minutes or more, stirring to thoroughly combine.

4. Run fruit mixture through a blender or food processor until perfectly smooth.

5. Pour fruit purée into the leather pan of your dehydrator in a layer ¼ inch thick. Dry in the dehydrator at about 140°F until you have a fruity roll-up texture, which will probably take 3 to 8 hours, depending on your model and the thickness of your fruit purée. Test frequently to prevent overcooking.

6. Store fruit leathers rolled up on plastic wrap or wax paper. Store in airtight bags or totes in the pantry and use as a sweet snack when ripe fruit isn't available.

Making Jerky in a Dehydrator

Before refrigeration, jerky was a method of preserving meat when an animal was butchered, so that nothing went to waste. In the wilderness you could eat as much of the meat freshly cooked as possible, and then smoke-dry the rest of it in thin strips. The dehydrated meat would be less likely to get spoiled and go bad so soon, and allow several weeks of use. These thin strips are also easier to carry because of the evaporated water.

Now it's not needed for long-term survival, but making jerky is a boon for the backyard farmer with a dehydrator. So pricy to purchase from the store, even in small portions, yet it's so easy to make on your own. My father-in-law makes his in the woodstove oven but it's even easier for beginners, and with better results, in a dehydrator. For people with a sensitivity to MSG, homemade jerky is the cheapest way to make sure your jerky is free of MSG and other unwanted ingredients.

In order to make jerky you need raw, lean meat (beef, venison, elk, goat, and other meats have been used successfully) that is sliced into 1-inch strips about $^3/_{16}$ inch thick. Trim any fat pieces off the meat and season the meat using brine or dry rub. Fat won't dry properly and will retain moisture, so be sure you remove it all if possible.

If you use a dry rub for the beef jerky you can use seasoning such as salt, pepper, garlic powder, and onion powder. Salt is the most important ingredient because in addition to helping remove the moisture, the salt acts as a preservative in keeping the meat from spoiling. Generally about ¼ of the overall seasonings used should be salt, but thicker-cut pieces will need more salt.

Coat the meat with the rub and let it sit in the fridge overnight. Place the salted strips in the dehydrator for the recommended amount of time suggested in your dehydrator's manual. Check at regular intervals from the lowest timeframe—you want the jerky dry, but still pliable, not crunchy and brittle. You know your jerky is done when it cracks but doesn't break when you bend it in half. Store in the fridge away from moisture for three months in an airtight container, plastic wrap, or vacuum-sealed plastic bags.

You can use inferior cuts of meat to make jerky, but you have to grind them first. Mix the ground meat with the seasonings and then roll it out into a flat shape. Dry it in the dehydrator and slice it into strips when it's finished. This method will take less time to season and dry.

MAKE YOUR OWN MARINADE

My favorite method of making jerky is using a marinade to cure the meat before drying it. One of my favorites is a simple mix of soy or Worcestershire sauce, brown sugar, garlic and onion powder, water, and a dash of liquid smoke. Experiment by changing the basic ingredients. Create an oriental flavor with soy sauce, ginger, and orange juice, or go Western with ketchup, red chili pepper flakes, and brown sugar.

Soak the strips of meat in the marinade in the refrigerator for at least 12 hours before putting the meat into the dehydrator. Watch the drying time carefully—if your dehydrator calls for 8 to 12 hours drying time, start checking the meat every hour beginning at 8 hours.

Making your own marinade instead of using store-bought marinades saves you money and allows you to control the ingredients so you can avoid MSG, artificial preservatives, or ingredients you don't want.

Drying in a Solar Oven

A fun project for the homestead is to build a solar oven to use instead of an electric dehydrator. Being able to harness the energy from the sun is a much more energy-efficient way to dehydrate your food than running a dehydrator all day long. The trickiest part is temperature regulation.

A solar oven works almost like a cold box. A clear glass top allows the heat of the sun to enter the box. Many solar ovens also have reflective floors to create radiant heat as well.

Many people prefer to supplement their seasonal produce drying by using the solar oven to cut back on energy costs, noise, and free up counter space indoors. And some homesteaders do all, or almost all, of their dehydrating using their outdoor solar ovens! Who knows, you might try it and find yourself hooked on the traditional and old-fashioned technique.

This solar oven can be used to dehydrate herbs, create jerky, and even bake foods, making it a great energy-free cooking alternative.
(Photo courtesy of Erik Burton)

A solar oven hot box design needs a clear top to collect the heat of the sun. It will need a deep-enough box area to hold the food items being heated and vent holes in the sides of the box to allow moisture to escape. Be sure to screen these vents so bugs and pests can't get inside and get to your food!

There are as many ways to build a solar oven as there are homesteaders. Most use whatever products they have on hand. If you want a place to start, though, there are full plans for a solar oven at BackyardFarmingGuide.com, courtesy of the University of Tennessee.

It's easy to use a solar oven. Set your solar oven where the window is facing toward the south in a spot where the sun will shine directly on it. Put in your herbs, vegetables, or other food that you wish to dry after you prepare it.

As always, it's best to preserve food that is at its very freshest. The sooner you preserve foods after harvesting the more nutrients and flavor you will maintain. That's the benefit of preserving your own food—you can pick your vegetables, fruits, and herbs at their peak.

Let your solar oven "cook" the food for a full day. In periods of hot summer heat check your herbs before the peak of the day, as herbs do not take as much heat to dry as some of the fruits and vegetables. Small, leafy herbs don't take as long to dry as larger produce like peach slices or tomato halves.

Making Sun-Dried Tomatoes

One of my favorite must-try fruits for a solar oven is the humble tomato. Sun-dried tomatoes usually take more than a single day in the solar oven and finish roasting halfway through the second day where I live. In cooler climates it could take up to four days. Sun-dried tomatoes can be very pricy to purchase from the store but so easy to make at home.

To make sun-dried tomatoes you can cut small cherry tomatoes in half. If you sun-dry larger tomatoes try slicing them into slices about ½ inch wide. Coat the tomatoes with any seasonings you want. We often use oil with basil to spread over the tomatoes. Spread them out in a single layer and place them in the solar oven to dry.

If you see any signs of moisture, you'll need to open the vents more to allow better airflow. Check your tomatoes as they dry and if you see any signs of mold or mildew, remove those pieces immediately and toss them in the compost bin. When your tomatoes are nicely finished, you can store in an airtight mason jar or zipper-lock storage bag.

Drying Meat in a Solar Oven

You can dry meat like jerky in a solar oven as well but I recommend adding a thermometer into the oven so you can watch the temperature levels. The USDA recommended minimum temperatures are 160°F for beef and 165°F for poultry. Without heating the meat strips to this temperature there is a risk of bacterial growth. Otherwise follow the same methods you use for making jerky in the dehydrator and enjoy energy-free cooking.

Spread your jerky meat on a dehydrator rack, oven rack, or solar oven to be dehydrated.
(Photo courtesy of Andrea Nguyen, Vietworldkitchen.com)

Smoking Meat in Home Smokers

Smoking is another low-heat method of preparing meat but instead of drying the meat, as with dehydration, or cooking the meat directly, as with traditional cooking methods, smoking heats the air and cooks the meat indirectly. This method takes much longer to cook the meat. But the meat is enhanced with a smoky flavor during the cooking process, and usually becomes very tender because it cooks at lower temperatures.

One of my favorite childhood books, *Little House on the Prairie,* had a chapter about a homemade smoker Pa made out of a hollowed log. The meat was hung from nails at the top of the hollow tree, and the fire kept burning and smoking at the bottom of the hollow tree. Thankfully, now we have extremely affordable home smokers that don't require you to hollow out a standing tree trunk.

The most typical home smoker is a water smoker that has a place for the heat source, racks to place the meat on, and a water pan to create steam and keep the meat moist. You want a smoker with a little door to allow you to add more fuel (charcoal or wood chips usually) without opening the entire lid. Incidentally, this is why trying to smoke meat in a regular grill doesn't tend to work as well—too much heat escapes each time you open the lid and the cooking temperature doesn't have a chance to stay consistent.

I think every home cook has secret ingredients when it comes to the perfect fuel and perfect wood type. In fact, one of the benefits of smoking your own meat is so you can control what type of chemicals your meat is exposed to. There's no reason to use charcoal with starter chemicals added to them, and certainly be careful not to use gasoline or other homemade starter fluids to kindle the fire.

Different kinds of wood chips and wood chunks create different flavors of smoke. Common woods that are good for smoking are hickory, maple, mesquite, or other hard woods. Do not use soft woods such as pine, because they produce creosote and aren't as suitable for high-quality smoking. Creosote will give the meat a bitter taste and actually numb your tongue when you eat the meat. Creosote is also a fire hazard if it builds up in the smoker, which is an important reason to keep the smoker clean.

Start your fire and have water in the pan. Some chefs like to put wine or beer in the water pan to enhance the flavor of the meat. The meat you add to the smoker should be thawed, as the low temperatures of the smoker won't thaw the meat quickly enough to be safe. Your smoker temperature should stay above 225°F to 250°F. The meat itself should be cooked thoroughly until a meat thermometer registers at least 145°F to 150°F for beef, venison, and pork roasts. According to the USDA, ground meats should be cooked to a temperature of 160°F, while poultry should be cooked to 165°F.

As with jerky making, you can marinate meat in a refrigerator for 8 to 12 hours before smoking. One of the most well-known cures for meat may be the classic pineapple and brown sugar, or salt and brown sugar treatments for smoked ham. Whether you decide to marinate the meat before smoking or not, remember to use common safety guidelines. Keep your meat cool in the fridge until you're ready to heat it.

You can smoke a wide variety of meats. Even meats that are hard to grill, like fish fillets and chicken legs, can be smoked. Most common cuts, however, are items like brisket, ribs, and pork hams. Experiment with changing the flavor of the meat cures as well as the type of wood you use to see what you like best. After you start smoking your own meat, you'll probably never buy BBQ meat in a restaurant again.

Root Cellars and Basements  19

Dating back thousands of years, farmers and families alike have preserved their harvest in the cool, dry climate of a cellar or basement. "Cellar," being used loosely, as earliest versions involved simply burying the produce until it was time to use it. The methods have matured over time, but the concept is the same. Keep produce cool and dry, and its shelf life will repay you in dividends.

Early American settlers had the hang of cellar storage well, with root cellars found in the thousands. One small town alone boasts more than 100. That a number of them in this so-called "Root Cellar Capital of the World" have remained for a couple hundred years is proof of their longevity.

You won't need a 200-year-old cellar, but you can still construct one that is practical and functional for your backyard farm. The benefits of easy storage to stretch your harvest will stand up against the initial work and investment.

Storing Crops in a Cellar

Consider when and how you harvest your crops. The condition of the crop itself is important because it will contribute to the quality and duration of storage. As soon as stems are broken and the environment changed, a crop will start to decompose. Keeping it in its environment for as long as possible will prolong its time in storage. Make sure that you are harvesting properly and at the right time for that particular crop.

Root vegetables can stay in the ground if you mulch straw over the top of it. Until the ground freezes, they can stay there for as long as a month. Onions are the exception, as they need to be harvested as soon as the tops fall over so that you can dry them out.

Pumpkin and other winter squashes can be left on the vine until the vine itself is withered from frost. In southern climates you would harvest when the stem begins to shrivel. Breaking the stem breeds opportunity for disease, so there is no reason to even touch it until you are ready to bring it in or until the vine has seen better days. Kale and collards can stay out in the garden past frosts, but spinach or romaine cannot. Find the optimal time to harvest each plant and wait until then.

When onion tops turn brown and begin to dry, you know they are ready to harvest.
(Photo courtesy of Tim Sackton)

THORNY MATTERS

Apples and some other fruits emit an ethylene gas that can make your other vegetables, such as potatoes, age faster, cause sprouting or rotting tomatoes, and make carrots bitter. In the small quantities of a backyard farm, you can probably avoid this by simply keeping the fruits closer to the ventilation so that the fumes are sucked away from the other food.

Crops can be picky things. The longer you intend to keep them, the more you will need to attend to their environment. Including a humidity gauge, or *hygrometer,* in your root cellar will help you optimize the space to accommodate as many of your fruits, veggies, nuts, seeds, and grains as possible.

Storing Root Cellared Foods

There's a return to root cellaring and other homestead arts for a good reason. The savings, self-sufficiency, and health benefits of fresh eating through the winter months make it easy to try this for yourself. You can get excess crops for next to nothing at the peak of the season, and then store these foods in the root cellar for weeks, months, or even longer.

Root vegetables. I think the image most people are familiar with when it comes to storing in a cellar or basement is of root crops. Beets, carrots, garlic, onions, and potatoes all store really well. We've eaten fresh potatoes all the way through winter and into early spring with no loss of flavor or freshness!

Potatoes should be cured for a few days after harvesting to allow the skins to dry out and toughen. Cover them with straw so the sun doesn't shine on them when you do so, or the light will make the potatoes turn green. The green bits on potatoes are toxic and should be avoided. Alternatively, find a warm, dark room to lay them out for curing out of the sunlight. When they are ready, they will store well in burlap sacks or in milk crates filled with straw.

Just these easy-to-keep vegetables alone would provide plenty of fresh food through the winter until the spring growing season starts up again. Onions and garlic are the only ones of these crops that prefer drier conditions; the others like cool and humid. Given the right conditions, some easy keepers like winter squash, potatoes, and apples will last for months.

OVER THE GARDEN FENCE

Be sure that the fruits and vegetables you store are in excellent shape. Fruits or vegetables with broken skin, blemishes, missing stems, or other flaws should be used right away. These would also be the fruits you want to use for fresh eating sauces, or cider.

Fruits. Don't think that fruits and berries can be stored through the winter? Think again! While many berries are fragile and fleeting (think raspberries), there are other berries and fruits that will stay fresh for weeks in the cellar.

Apples, pears, and grapes will all store in cold and moist conditions. Give them a little bit of humidity and not-quite freezing temperatures and you'll have the ability to eat fresh fruit for weeks. Remember, store the fruits away from the other vegetables, as they can speed overripening of your veggies. Check over your fruit every week or two to make sure you are removing any spoiled fruit.

All of these fruits will store best when they are not overcrowded and directly touching. We use a milk crate to store our apples, for example, and avoid spoilage by laying straw between each layer of apples. Spacing them far enough apart so there is a gap between the skins of individual fruit helps as well. Fill the milk crate, layering fruit-straw-fruit-straw, and then carefully store in the cellar away from the potatoes. Alternatively, you can wrap the apples, pears, or other fruit in plain newspaper so the paper prevents moisture build-up and rotting.

Beans and peas. Your legumes will want a cool and dry place, as humidity can cause the dry beans to try to sprout. Stored in an airtight container, dried beans, peas, and other legumes will store for years. Be sure that they are completely dry before storing.

Squash. Squash plants are a well-known crop for root cellars. Winter squash have thicker skins that store well during the winter. Storage is best when you let the squash fully ripen before harvesting and leave some of the stem attached. Let the squash sit in the sun for a few days to help toughen the skin, a process called *sun curing*. Turn the squash a quarter turn each day so the sun reaches all sides of the squash, and it will harden evenly. If you aren't able to sun-cure the squash, store them at room temperature indoors for two weeks, rotating daily as described previously.

Often in the peak of season you can find great deals on large lots of fruit and vegetables. Storing these in a root cellar will provide months of fresh fruit.
(Photo courtesy of Cindy Funk)

The winter squashes in this crate are all suitable for storing in the root cellar.
(Photo courtesy of Jen Dickert at SuperNaturalNutrition.net)

You may have seen pictures of a heap of pumpkins in the corner of the root cellar. In reality, if you store winter squash so they are touching, the skins will soften where they are making contact. Don't stack them on top of each other. Keep them insulated with a pile of straw between them.

Stored properly, these winter squashes can last up to six months, just about the time the thin-skinned summer squashes begin to ripen in the garden. Be sure to keep the stems intact on your squash. If you cut them too close you'll make it easy for bacteria to get in and make the squash rot.

Cabbage. This good fall crop can store well for several months. Harvest your cabbage for storage as soon as the heads tighten up, and remove any loose leaves. Cut off the stem completely, and you can wrap each in butcher paper. When they are wrapped, you can store them in a wooden box or milk crate filled with straw, loose sand, or sawdust.

Cabbage stores best in tight heads like these.
(Photo courtesy of Tim Sackton)

OVER THE GARDEN FENCE

Wash your stored fruits and vegetables before use, not before storage. Just shake off the loose dirt and debris; washing them in water before storage can lessen the amount of time you get in storage. When you bring the crops up to the kitchen to use them, that's the time you give them a thorough washing.

Constructing a Simple Root Cellar

Storage in a root cellar is simple, but it is only successful if you keep some conditions in mind. The initial setup takes some forethought; from there it is easy to maintain and utilize.

All root cellars need just a few common elements to be successful. They should be dark. They need to stay humid and cool. And they should be heavily insulated, the most common method of insulation being several inches of soil. Your root cellar can be as elaborate as a full-sized room dug into the side of a hill, or as simple as a trash can buried in the ground.

Trash Can Root Cellar

This simple technique allows you to create a mini-root cellar if you don't have a basement or large root cellar. It will also allow you to store different crop items apart from each other. For instance, you can keep your apples away from your potatoes so everyone will be happy. Best of all, this root cellar system can be done in a single day, so you'll be able to capitalize on an unexpected bounty or awesome special you found on bulk veggies at the farmer's market.

To build your trash can root cellar, you need a large trash can with a lid. Don't worry about paying extra for the cans with wheels because you'll just bury the trash can anyway. It doesn't matter whether the can is metal or plastic—I've seen these built with both types.

Dig a large hole in the ground in an out-of-the-way location where you won't be walking directly over the area. You want the trash can to fit well into the hole except for a little gap directly underneath the trash can to allow moisture or condensation to drain. You can use a drill to create a couple small holes in the bottom of the trash can to allow any collected water to drain away. The very top and lid of the trash can will be just above ground level.

After the trash can is buried, you can line the bottom with straw. Then add your veggies or fruits in milk crates filled with straw or sawdust to keep them insulated. Stack the baskets of crops on top of each other until the can is full. You can place extra straw around the gaps in the sides, but the main insulation for your veggies will come from the soil around the trash can.

Place the lid securely on the trash can and cover over the top of it with straw. If you live in an area with a damp winter, you could cover the top of the lid and straw with a tarp secured with rocks or bricks.

Check your produce on a regular basis as you would with any root cellar. Look for veggies or fruits that are getting soft, or showing signs of mold or rot. These should be removed immediately and eaten if salvageable or added to the compost bin otherwise.

Basement Root Cellar

The most common and familiar setup for a root cellar is a full room underground. Often, this is a basement area under the house itself. Sometimes it can be a simple dug-out room at the base of a hill outside the house. Both provide for the crops to be stored in a cool location that is insulated during the winters from frosts and freezes.

If you have a finished basement that stays relatively warm (more than 50°F), then you may need to create a root cellar corner in your basement by walling off a section of your basement and not ducting in the central heat. Include a small vent to the outdoors which you can open each evening to let the cold air come in and help maintain a cool temperature just cooler than 40°F.

Digging out a small root cellar along a hill or slope in your backyard may require a building permit depending on the regulations in your area. Also, be sure to check with your utility companies to make sure you don't damage any underground wires or pipes. The nice thing about a hillside is that you don't have to dig out a full shed-size hole. You can dig it out partway, and then extend the hillside over the shed structure with soil several inches thick to allow it to stay insulated.

If you live in an arid area and need to increase the humidity of your root cellar, you can place a dish of water on the floor. I've also heard of people placing damp burlap sacks over the milk crate or bucket full of veggies and hay. The hotter your climate gets, the deeper or better insulated your root cellar will need to be.

Preserving Garden Herbs

20

Herbs serve such amazing purposes throughout their lifecycle. Seeds not only propagate new plants but can often provide culinary and health benefits, as with coriander (cilantro) seeds, for example. Inside the garden, herbs make fantastic companion plants, warding off harmful bugs or attracting helpful ones as needed. And of course, leaves and flowers can be harvested throughout the lifecycle to be used fresh or dried, in tinctures and teas, salves and pastes.

Some herb and vegetable pairings will actually improve each other's flavor, as with basil and tomato. Some have culinary and medical purposes both, such as rosemary, which is not only popular for seasoning potatoes, but is also an excellent scalp and hair conditioner, as well as memory and concentration enhancer.

Growing Your Own Herbs

Even the smallest of herb gardens can yield long-term benefits if you preserve the harvest with uses in mind. While so many people buy herbs as they need them for cooking and then toss what gets left behind, you can employ your steady harvest to use in many ways for the months and even years ahead. And the bounty can be shared or even marketed and sold.

THORNY MATTERS

Herbs have been used medicinally for centuries, and modern scientific research confirms the suitability of many herbs when used in medicinal quantities. However, you should use herbs medicinally only under the supervision of a trained health-care professional. When we agree that herbs can help benefit our body in certain ways, we are saying that they change and affect our bodies. That effect is not always a good one in all people. For example, elderberry helps boost the immune system and fights against viruses like the flu and the common cold. However, it can also stimulate the immune system in people with an autoimmune disease, and so should be avoided by those people.

Containers are often ideal for herbs because many of them will take over your garden if allowed. As a bonus, this makes them easier to maintain and you can keep them near the door for literal easy picking.

Drying Herbs for Teas

In Chapter 18, I talked about dehydrating and drying herbs. But what do you do with them after they are dried? You could use many dried herbs to season food, but their benefits and flavor are absorbed and enjoyed thoroughly when made into teas. And let's be honest—there's nothing quite so satisfying as curling up with a book and a hot cup of tea at the end of a busy day.

Herbal teas should be steeped in already-boiling water for at least five minutes before drinking. If you don't have anything to steep with, you can pour the water over loose leaves and strain it or drink as is. However, there are plenty of gadgets available in stores to put the leaves in and let the water steep through and around it without having a leaf-filled tea.

The simplest solution is to simply roll a bit of your dried herbal combination into a double layer of 6-inch square cheesecloth, then fold the ends up and stitch and tie off with some string or embroidery thread. For larger batches of tea, you can use clean pantyhose, like the 50¢ single knee-highs they sell in little plastic eggs.

Herbal Tea Benefits

The immediate benefit of sipping herbal tea is that it could be much worse! Caffeinated or sugar-laden drinks are not typically helpful to you and can prove addictive. You might get hooked on your favorite tea, but it won't be the same as the drive to have coffee or the possible headaches when you skip soda. Even replacing one cup of soda with hot or iced herbal tea helps you avoid the drawbacks of soda while also supplying the beneficial health properties of the herbs. It's a total win-win.

Additionally, drinking a tea made from your own herbs will help you absorb the nutrients that you might miss in a supplement. It's very hard for your body to break down a pill, especially a tablet. Sipping a liquid is much more readily absorbed, especially when it occurs in nature to provide just the right combination. For example, iron supplements are often poorly absorbed in the body, leading to digestive troubles. On the other hand, the iron in a tea made from nettles and parsley is naturally combined with vitamin C, meaning you get the most iron it has to offer, plus the boost you need to use it.

Herbs have been studied and used for their various benefits for millennia. Each plant has its own special attributes, and combining them into teas gives you the chance to make your own holistic approach. Because you are growing the herbs yourself, you also have careful control over what exactly you are harvesting and what has touched the plant over its lifetime. No worries about pesticides or mistaking plants here—that is, as long as you label your seedlings!

Common Herbal Tea Blends

Iron-boosting tea is common in my circle of moms. Nettles, which you must harvest carefully before drying to remove the stinging effect, parsley, and mint combine to make a powerhouse iron booster, but with a softened flavor from the mint. Pregnant women often add red raspberry leaf tea to further the iron and calcium benefits and make a well-rounded tea for pregnancy health.

Chamomile is a staple of relaxation tea, often used alone or with a bit of fruit, like orange zest or apple slices. It can also be combined into other medicinal or fruity blends to bring a calming factor and a bit of sweetness.

Rose petals can be brewed fresh or dried, just cut the white part off the bottom of the petal to remove any bitterness. Fragrant roses will taste best, and their high antioxidant levels make them a good addition to many tea blends. Hibiscus blooms also provide antioxidants, as well as potential benefits against diabetes. With a bit of a tart flavor, you may want to sweeten it or combine with other flavors to soften.

Ginger is a good addition to a tea for soothing digestion. Peppermint accomplishes the same, and can be added to nearly any blend to provide a cooling flavor. Strawberry leaves, lemon grass, and stevia all add sweet or fruity flavors that can be used to increase the flavor of tea depending on what you enjoy.

One of my favorite blends when I was pregnant was nettles, lemon grass, red raspberry leaf, and mint. Red raspberry leaf tea is a known uterine tonic, and the other ingredients gave it great flavor and impart important nutrients like iron and vitamin C. Have fun experimenting and see which flavors you like best.

Hibiscus flowers are absolutely gorgeous in the garden, great for teas, and edible as well. So the same flower could attract bees to your vegetables, decorate a birthday cake, and simmer in your tea.

OVER THE GARDEN FENCE

The longer you brew the tea, the more benefits you'll get from the herbs but the stronger the taste will be.

Herbal Vinegars

Vinegar is made from fermented natural sugars, hence the variety of "real" vinegars—white wine, red wine, apple cider, rice—coming from starchy foods. Even balsamic is made from grapes and then aged for years. White, distilled vinegar is mostly a lab creation and better suited to house cleaning and pickling.

Real vinegar will have flavor subtleties and health benefits that the distilled stuff has lost. The claims include everything from kidney health to blood sugar and blood pressure improvement. Now, add in herbs and you have a powerhouse of flavor and health.

To make herbed vinegar, simply lay herbs that have been washed and patted dry into a sterilized jar. One cup of dried herbs combines well with about three cups of vinegar. Pour the vinegar over it, cover with a nonreactive lid, and let sit in a cool, dark place for a few weeks. Check back to see how the flavor and aroma are. When it is as strong or subtle as you'd like, strain the herbs off and pour it into a fresh bottle or jar, maybe with a couple of sprigs for decoration. Don't forget to label and date your jars so you know what they are.

Flavor combinations are endless. White vinegars can be tinted with blossoms and certain herbs, like purple basil. Red vinegars hold up well against strongly flavored herbs. Beyond that, simply use the herbs you like. Lavender blossoms are wonderful in white vinegar. Other basic combinations might be rosemary and garlic in white vinegar, raspberry and black walnut in red wine vinegar, or tarragon with thyme and garlic in cider vinegar.

Basil has such versatile flavor that it can be added to nearly anything. In white vinegars, it will leave a pink tint, which adds aesthetic appeal to an already attractive infusion. This basil has already begun flowering, which makes a stronger flavor in the leaves.
(Photo courtesy of Steve Swayne)

Experiment with blossoms, herbs, fruit and citrus zest, nuts, seeds, and even peppers. Vinegar is also a good rinse for your hair and clothes, so herbed vinegars are even better. Of course, you might want to skip the garlic and onions if you plan to pour it on your hair or rinse your mouth out with it!

If you enjoy making new flavors and have the harvest excess to indulge the hobby, you will probably find yourself with more herbed vinegars than you know what to do with. They are so simple to make and so fun to play with. Find old bottles and jars, especially ones with skinny necks that you can cork. Clean and sterilize them, then fill with your herbed vinegar with a beautiful sprig of parsley, lavender, or any other herb. Make a nice tag tied around the bottle with string, and you can even decorate the bottle and cork or lid if you have an artistic hand. Not only do these make lovely gifts, but they are quite marketable if you would like to sell them. With such a low level of work needed to make them, even a seemingly low price can be worth the effort.

Herbed Butters, Oils, and Spreads

Think for a second about how you use butter. Whether it's sweet or savory, how could it not benefit from some extra flavor? Lavender and honey butter on toast in the mornings. Rosemary, basil, garlic butter on dinner rolls in the evening. Now think about the oils you cook with, sauté veggies in, drizzle over salads. And mayo or cheese spreads, sprinkled with peppers and cilantro or oregano and parsley.

Herbed mayo makes simple sandwiches taste like gourmet creations.
(Photo courtesy of Rachel Matthews)

Mayonnaise, butter, and cheese (such as ricotta or cottage cheese) are quite simple to combine with herbs. Simply make your herb mixture with dried or fresh herbs, then stir it into the softened butter or other spread. Remember that 1 teaspoon of dried herbs is equivalent to 1 tablespoon of fresh. You can then store them as normal in the fridge and they should keep for the same length of time as condiments without herbs added.

Oils, on the other hand, can be a breeding ground for botulism if not attended to, so you must exercise a bit more caution. Combine herbs and oil by mixing your blend in a slow sauté for a few minutes until warm. Only make about a cup at a time, which will use about ½ cup of fresh herbs.

This is because you must not store the oils long. After you have strained the oil, store it in the fridge for just a few weeks and use it liberally.

Incidentally, garlic oil made this same way and cooled to room/body temperature is great for soothing earaches.

Herbal Infusions, Decoctions, and Tinctures

The ability to make your own home herbal preparations will not only provide another way to use and preserve what you've grown, but will also save you money. If you've ever purchased an herbal remedy you know they can be very pricy. Making your own echinacea extract or calendula salve can increase your overall wellness while saving you money by not having to purchase expensive herbal extracts. And of course, you'll be confident about the quality of the herbs and plants being used because they are your herbs and plants.

One of the easiest herbal home remedies to make is an herbal *infusion,* which is almost like a really strong tea. An infusion is especially useful for softer herb parts like flowers and tender leaves. Place 2 cups of chopped fresh herbs into a cleaned and sterilized quart jar (the finer chopped the herb, the better the results will be) while you bring a saucepan of water to a boil on the stove. When the water is boiling, pour it over the herbs in the jar until the jar is full. Put the lid on loosely and let the herbs steep for 8 to 12 hours. Strain the liquid through a filter or cloth to extract the infused water and bottle it up. The leftover herbs can be tossed in the compost bin.

These water-based preparations are much stronger than a simple tea, but are potent, healthful drinks. They will not last as long as an alcohol-based preparation, so make them as you need them. Usually an average adult would only drink 1 to 3 cups of herbal infusion each day, but there are other uses for the herbal infused water. You can use some infusions as hair rinses, in bath and body products, as substitutes for plain water in soap making or lotion blends, and more.

One infusion I use on a regular basis is a rosemary infusion. I bring the jar with the rosemary-infused water into the bathroom and use a cup as a hair rinse in the shower. It's stimulating to your mind, though, as well as adding shine to your hair, so only use it in morning showers and not before bed.

A *decoction* is a much stronger brew because the herbs are placed into the boiling water and the liquid is allowed to simmer down until it's much more concentrated. The total liquid volume is ultimately decreased by one fourth. I rarely use decoctions unless the taste of the herb is so strong or unfavorable to me that I wouldn't want to drink it as a tea or infusion. Water-based decoctions won't last long, so these are other herbal preparations that you should make as you need them.

Tinctures are traditionally alcohol-based, although some formulations for children are sold as glycerin-based tinctures. These are more properly known as glycerites even though the preparation methods are basically the same. True tinctures are alcohol- or water-and-alcohol–based which ensures that the solvent (the base) absorbs as much of the medicinal properties of the herb as possible. This alcohol-based liquid has a highly concentrated amount of the herb properties steeped into an alcohol base.

DEFINITION

An **infusion** is made by pouring boiling water over the herb material and allowing it to steep for a time (like a tea). Some infusions are made with oil instead of water. A **decoction** is made by placing herbs into boiling water and simmering them for about an hour (time varies). The herbs are then strained away and the water stored and used. A **tincture** is used when the plant's substances don't dissolve as readily in water and is made by covering the chopped-up herbs with alcohol and allowing them to soak for several weeks. The herbs are then strained out.

Some people are concerned about the alcohol in tinctures. However, the amount consumed in most doses is so small that it doesn't impair you in any way, but rather serves as a preservative base for the healing properties of the herb.

The method I'm sharing today is more of a "folk" method and doesn't rely on the careful measuring and weighing that a professional herbalist would use. Without measuring techniques to ensure uniformity between batches, these tinctures won't be suitable for sale but will enhance the overall health of your family.

To create an herbal tincture, follow these steps:

1. Chop up fresh herbs or coarsely grind your dried herbs. Put the herbs in a large jar— canning jars work well for this. With fresh herb material you want to fill the jar as completely as possible.

2. Add 80 or 100 proof alcohol to the jar and fill it completely so all the herbs are covered by the alcohol. Put the cap on tightly but make sure you wipe the rim completely clean first so there is a good seal.

3. Now the fun part! Every day, shake the jar vigorously to help speed the infusion of the alcohol with the herbal benefits. We sing silly songs and dance about the room wielding the jars of macerated herbs like maracas. Fourteen days is a good minimum for this process but if you forget for a few days and your herbs stay in the alcohol longer, no harm done. Some herbal materials take longer.

4. Strain the herbs through a filter, catching the richly infused alcohol in a clean container. Squeeze out the leftover herbs so you get every bit of juice out of them and then discard them into the compost bin. The herbal tincture can be bottled into a glass vial with a dropper top for easy administration.

Tinctures are more concentrated than infusions, which are in turn more concentrated than tea. If tea is consumed by the kettleful, and infusions taken by the cupful, then tinctures are taken one dropper-full at a time.

Tinctures will store a long time, often indefinitely, because the high amount of alcohol acts as a preservative. Just shake up the vial before administering to make sure it's mixed up.

Sometimes different herbs do well prepared different ways. Peppermint makes a nice tincture from dry leaves, but with fresh leaves on hand, try infusions. Echinacea does well as a tincture as well because to me the taste is not pleasant for a tea or infusion. Nettles, on the other hand, are extremely flexible and widely used in teas, infusions, decoctions, and tinctures.

For more information on making infusions, decoctions, and tinctures, as well as the other topics in this chapter, a good resource is *Making Herbal Medicines (A Living Free Guide)* by Susan Mead.

While we visit our medical care professional on a regular basis, and are not opposed to medical treatment when necessary, there is a huge benefit to understanding how plants can help improve your overall health. Better nutrition, more awareness of healthy living, and overall wellness can certainly be enhanced by adding more whole foods to your life, and these easy herbal preparations are just a part of that.

Crafting from the Backyard Farm 5

This part recaptures the lost arts of homesteading. It's the perfect part to flip through hither and yon on an as-needed basis, or read straight through to pique your interest in what is possible. You'll be tempted to join the renaissance of those reviving the do-it-yourself concept and relish in the satisfaction of a job self-done.

Learn how to use the milk from your own dairy herd in soap making, cheese making, and other useful ways with the shared recipes and techniques. Brew homemade hard ciders or wine from your homegrown fruits. Dye the fibers you've spun from your sheep, rabbits, or goats.

These are the skills you grew up hearing your grandparents talk about doing. And more and more backyard farmers are finding pleasure, and profit, in revisiting these skills for themselves. After browsing these chapters you just might find the same.

Making Butter, Yogurt, and Cheese

 21

Dairy foods are largely equated with indulgence. (Deep-fried cheese, anyone?) Over time, they have been alternatively vilified and glorified. The deep-fried cheese probably has something to do with that. Just a thought.

For all of the controversy, the health benefits of live cultures that you get in homemade butter, yogurt, and cheeses are undeniable. If you have incorporated a dairy animal into your backyard farm, you understand the benefits that dairy can provide, particularly when it is safely unpasteurized and cultured.

Butter

One of the simplest ways to reap these benefits is, in fact, with butter. Contrary to mainstream belief that has us drinking skim milk, the cream is the best part. It carries fats and vitamins that are essential to gut health, brain function, and the immune system, among other things. And coming from raw milk, none of the proteins that deliver these nutritional giants have been destroyed.

Even if it has been pasteurized, you might be able to find milk in the store that has not been homogenized. That is the process of straining milk through small tubes to break down the fat molecules so that they mix evenly through the milk rather than separating to the top. This separation is key to butter making, and without it you would need to purchase cream as your starter.

Butter from Raw Cow's Milk

Straight out of the cow, your milk will separate in the fridge and set you on your way to butter making. We sometimes buy our milk straight from the Amish dairy the day it's milked, and it makes amazing butter. You may have to look around to find your own milk source but as you get into the world of self-sufficiency and backyard farming, you'll find an amazing community you never knew existed.

Cream itself is useful, but separated into butter and buttermilk, it will stretch much further.
(Photo courtesy of Brannan Sirratt)

OVER THE GARDEN FENCE

You may be able to buy a share of a milk cow, or participate in the dairy Community-Sustained Agriculture (CSA) to source your own raw milk more inexpensively than from a health-food store. If you don't have room for your own milk cow, the next best thing to do is support a local organic farmer in their endeavors!

And chances are someone in that community will have fresh milk available for you to use in your butter and yogurt making. After the cream rises to the top of your chilled milk, "skim" it off with a spoon. Do this carefully until you start to see a bit of milk in it as well. Culturing this cream—leaving it covered and on the counter for a few hours—will increase the health benefits as well as the flavor of your butter.

It will take about 12 hours for your cream to be ready. It should smell just a bit sour, and be roughly 75°F, or room temperature. How do you know if it's sour enough but not too much? The nose knows. In all seriousness, it is something you will develop a knack for over time. Fermentation manifests itself in a sour taste and smell, so there will be a bit of a tang to it. Trust your instinct to know if it smells bad and overdone.

Now, your cultured cream is ready to be whipped into shape. Put it into a jar or churner, but make sure it doesn't take up too much room or it will not agitate properly. Leave the majority of the container empty. And shake.

When the contents feel heavy and the churning feels different—you still aren't done. But close! The butter will turn yellow and granulated when it's ready. If you shake or churn much after that, it will wind up very hard in the end.

Butter separates from the buttermilk as it nears completion.
(Photo courtesy of Brannan Sirratt)

Pour the butter into a fine colander, let the buttermilk drain out into another container, and set it aside. This cultured milk has lots of health benefits as well, so you don't want it to go to waste. Rinse the butter that is left in the colander with cold water until it runs clear, and then move it to a bowl. Salt it to taste, and work the butter around to get the water out. Some buttermilk and water may remain, and that's okay.

Put your butter in a container, or jazz it up with some herbs, such as chives or rosemary, and garlic first. Back into the fridge it goes, though you can leave it out a bit to get it soft and spreadable.

Butter from Goat's Milk

If you are making your butter from goat's milk, you will find that the separation stage is not so simple. Even after a few days, the cream still might not have risen to the top. Goats make incredibly rich milk for their babies, which is good for the baby goats but not so good for the goat-milking, butter-making humans. If you can get your hands on a separator, you can separate the cream right after milking. Separators can be purchased for about $200 to $300 through mail-order supply stores.

If you do manage to get some goat milk cream and culture it, the next step is to heat the cream in a double boiler. As soon as it hits 146°F, cool it to the 50°F to 60°F range by setting the pan in cool water. It's now ready to be churned, a process that will likely take around half an hour. After churning it to pea-sized granules, pour off the liquid milk that is left to be used elsewhere, and pour in fresh water. Churn a bit more, drain, and refill. Do this a few times until your rinse water is clear.

Now, you can spread your goat butter into a dish, salt it to taste, and knead it with a spatula until you can't squeeze any more liquid out of it. It should conform into any shape you "knead" it to (Ha!), just don't forget to keep it in the fridge.

Leave your refrigerated goat milk uncovered until it chills. Warm goat milk will produce condensation on a lid, and that will affect the flavor as it drips back into the milk. Of course, you will want to cover anything else in the fridge so that the milk doesn't pick up those flavors. (Unless you like garlic goat milk!)

OVER THE GARDEN FENCE

Butter freezes well, so if you are making a large batch and don't plan to use it all, wrap it in plastic wrap, and then put it in the freezer for use at a later time. Butter will store for about five to nine months, and salted butter will last longer than unsalted butter because the salt acts like a preservative.

Tools You Can Use

At the most basic, all you need to churn butter is a jar big enough for your cream. Shaking the jar can be a family affair, one that even the littlest can enjoy. In fact, you could wrap the jar in towels or bubble wrap and have kids roll it back and forth on the floor until the butter forms.

Kids get a kick out of shaking a jar to make butter, and as a bonus, it gets some of their energy out, too!
(Photo courtesy of Brannan Sirratt)

A churn is not necessary to butter making, but it might make your life a bit easier if you plan to make a lot. Churns have taken many forms over the years. The most interesting might be the rocking chair churn, which no doubt made for some happy mommies as they rocked their babies to sleep without neglecting their butter-making chores. Even today, you can buy electric churns, hand-crank churns, plunger-style, dasher-style, cylinders, and more.

ON A DIFFERENT SCALE

Just experimenting with butter? Put your cream—even store bought—into the blender and let it go. After a few minutes, you can pour the buttermilk out from beneath it, salt and blend a bit more, then rinse under cold water while kneading and squeezing it together. Voilà! Yummy, fast butter that will whet your appetite.

A little creativity goes a long way. You can certainly think outside the box to get your butter made. Among the more intriguing ideas, I find the idea of a gallon paint mixer to be absolutely inspired. If you can get a container to fit inside it, this could be a really efficient and creative way to make butter!

Yogurt

Milk is a somewhat delicate product that can spoil quite easily, especially in the days when there were no refrigerators or preservatives. For that reason, it didn't take long for people to learn to control the "spoilage" via fermentation and use cultured milk to their advantage. And advantages there are! Not all bacteria are bad bacteria, you see, and some are actually vital to our well-being. Yogurt is an age-old example of this, and surprisingly simple to boot. Buttermilk is also a fermented/cultured milk product.

The difference between cultured products and simply milk-gone-bad is temperature and environment control. Introducing the right bacteria and then keeping it at a favorable temperature turns the sugars in milk (lactose) into acid (lactic acid). As the neutral milk turns into acidic yogurt, the milk-protein casein reacts and thickens it.

Because of this, the ingredients needed to make yogurt are simply milk and a starter. If you have not yet made yogurt or do not have access to someone who has, Dannon Plain is a good starter. Whatever brand you use, be sure it has active live cultures. Be careful to not open it until you are ready to use it. We only want to introduce the good bacteria—not any foreign intruders.

As for the equipment and tools you need to make yogurt, it depends on your method. Because temperature is such a vital factor, you should have a dairy or candy thermometer on hand regardless of the method you choose.

Before fermenting, combine the starter and milk. It doesn't matter for usefulness if you use whole milk, skim milk, or anything in between. Practically speaking, it would make sense to use the milk left after you skimmed the cream off for butter.

OVER THE GARDEN FENCE

Store-bought yogurt is congealed with pectin, so you probably won't be used to it being very thin at all. The texture will be a little different. If you prefer thicker yogurt, try adding in a few tablespoons of powdered milk. It should thicken the yogurt without changing your recipe or process at all.

To open up the milk proteins and heat away any unwanted bacteria, strongly consider scalding the milk. This is heating it in a heavy-bottomed pan or double boiler until it reaches 185°F to 195°F, then immediately submerging the pan in cold water to cool the milk down to 120°F to 130°F and no hotter. In effect, this is very similar to high temperature/short time pasteurization processes used by commercial dairies.

Cooled to this temperature, the milk can be combined with your yogurt starter. For 1 gallon of milk, you need 1 cup of yogurt. Mix 1 cup of milk with the starter first, and then slowly stir that mixture into the rest of the milk. At this point, the milk and yogurt blend is called *inoculated milk*.

Now you have to keep the yogurt at the proper temperature for the next several hours. Too hot, and the bacteria will be killed. Too cold, and it will not ferment properly. There are many ways of accomplishing this:

- **Slow cooker.** Actually, the entire yogurt-making process can happen in the slow cooker. You can heat it to scalding on high, then after cooling and inoculating with the starter, turn it off, unplug it, and cover with a beach towel so that it holds temperature overnight (12 hours).

- **Thermos/cooler.** Pour your inoculated milk into sterilized jars. Cover them, and place them in a cooler of water warmed to 120°F to 130°F. Or for a smaller amount, pour it into a thermos and wrap the thermos in towels. Leave it for several hours. After five or six hours, the yogurt should start to gel.

- **Sun.** If you have a spot where kids or animals will leave it alone, you could set your mixture in a covered glass dish in the sun. Keep an eye on it so that it is left alone and in direct sun the whole time. After five or six hours, it should be ready. But that is a long time to go undisturbed in the sun, and you should be careful on hot days that it doesn't get too warm.

- **Oven.** Preheat your oven to 100°F, then turn it off. Set your inoculated milk blend in a covered dish and in the warmed stove. Let it sit (with the oven closed) overnight, and you should have yogurt ready for breakfast. A slow cooker can be used as an alternative, or you can set the milk blend over a lit oven pilot light.

However you choose to make yogurt, the most important thing is to maintain temperature control until it has set. Towels are great insulators to help this process.
(Photo courtesy of Brannan Sirratt)

If you want a thicker, Greek-style yogurt, pour it into a cheesecloth or towel to drain off the whey. If the grain is too open in the fabric, it will not strain selectively enough. A simple, thin towel or handkerchief is just what you are looking for. Keep the whey for smoothies and cooking, though—that's good stuff!

Cheese

A very basic cheese is made by taking the Greek yogurt method a bit further. Drain for a full 24 hours, and then chill it in the fridge overnight. The texture should firm up so that you can mold it into a ball. Congratulations! You have made *labneh.*

> **DEFINITION**
>
> **Labneh** is a soft cheese made from strained yogurt. Sometimes it's called yogurt cheese or Greek yogurt.

There are so many different kinds of cheeses that it is hardly possible to give you a complete guide for making cheese here. Each type of cheese requires its own chemical reactions, technique, and time frame. It also takes a bit more precision to make cheese than it does to make butter or yogurt. For example, while you can use pasteurized and ultra-pasteurized milk for yogurt, those processes change the structure of milk too much to allow for good flavor or general success. If you must use pasteurized milk, you will have to add calcium chloride to ensure that a good curd forms. Pasteurization neutralizes the calcium in milk too much for it to curd properly. (Doesn't that tell you something about the benefits of unpasteurized milk in general?) Raw milk is your best bet, though you do need to allow for extra curing time if you are unsure of the possibility of any pathogens in it at all.

Because cheese making is a kind of fermentation, you need to control the bacteria and temperature just as you do for yogurt. This requires a starter as yogurt does. Buttermilk and plain yogurt can work, but depending on the type of cheese you want to make, you might need to purchase one or grow a bacterial starter.

When the milk is acidic, the addition of rennet will change the consistency of the milk so that it becomes a thick gel. Remember the casein reaction when we made yogurt? Rennet will intensify that process even more. Casein, a milk protein, is water soluble. In acid, it is not. Because this transformation causes such thickening, you need to make sure your milk and starter do not thicken before the addition of rennet. Rennet can be purchased at grocery stores or from a cheese-making supply house. Natural rennet comes from rennin, an enzyme in a young animal's stomach that helps them digest their mother's milk. In theory, you can extract your own. It might be an interesting undertaking, but I'm thinking a quick stop by the grocery store or a call to a supply house will be just fine in most cases!

Inoculated milk with rennet stirred in should be left overnight, maintaining temperature control much like you do with yogurt. By morning, the consistency should be completely gelled. You are looking for what is called a "clean break." A finger pressed into it and lifted out should break the solids cleanly. What you have now are curds, and the liquid still in it is whey.

The milk should congeal to the point that it breaks cleanly around your finger.
(Photo courtesy of Brian Boucheron)

THORNY MATTERS

If milk has started to turn before you begin the process, the bacterial balance will be damaged and you may not get the chemical reaction you need for a clean break.

Hum the nursery rhyme and pretend you are Miss Muffet, because now the curds and whey must be cut. Each recipe will vary, telling you how to cut or stir the curds. Often, you will slice it with a long knife, making parallel lines until you have cubes of the directed size. Sometimes, you will be directed to simply stir them.

After straining the curds, retain the leftover whey to be used in other capacities.
(Photo courtesy of Brian Boucheron)

Next, the curds must be set. This is achieved by reheating them just a bit—the warmer, the firmer, but not much more than 100°F or you will kill the enzymes—and gently stirring by hand to ensure that it heats thoroughly without burning anything at the bottom. The warming will contract the curds, changing the consistency again.

When the desired consistency is reached, remove them from the heat, and the curds should sink into the whey. Now, they must be separated and salted. Salting the cheese inhibits the bacterial process and helps eliminate the moisture from the cheese. It is at this point that you have cottage cheese, and can reserve some before moving on with the rest of it.

Some cheeses can simply hang in cheesecloth to allow the whey to drip out. They will not be very firm, but they are good to begin with to get accustomed to the process. Others will need to be pressed. For $100 to $200, you can find a cheese press. Or you can fashion one yourself out of PVC pipe, a wooden block, and some other odds and ends. The goal of the press is to remove as much whey from the curds as possible. Firmer cheeses will require more pressure.

Time and pressure will condense your curds into a solid block of cheese.
(Photo courtesy of Brian Boucheron)

After a night hanging or pressed, some cheeses will be ready, and others will need to cure. This is often just a matter of wrapping with gauze "bandages" and leaving in the refrigerator as long as you can stand it, changing the wrapping as it dampens.

With all of that work behind you, the hardest step is the waiting. After a few weeks in the fridge, the outside should be hardened and yellowing. You could eat it now, but remember that cheese gets better with age. Because air causes mold to grow, you have to protect the cheese from drying and molding as it ages. To do this, melt wax over very low heat in a pan large enough for your cheese wheel to roll in it. When the wax is melted, very slowly and carefully roll the cheese in it to coat all sides. Dry (which happens quickly), and do another coat. Repeat until you have a smooth, waxed surface that you cannot see through. This will seal off the cheese as it sharpens, and maybe make it less of a temptation to nibble at!

Making your own butter, yogurt, and cheese is a truly satisfying craft. Not only is it beneficial for your health, but it is like making a gourmet dinner from the garden harvest. There is something very special about bringing all of that work into your kitchen and onto your table—especially if you haven't been raised on a farm or homestead environment. What once was limited to the supermarket, you have now made with your own hands.

The longer you cure cheese, the better it will be. That is, if you can wait at all.
(Photo courtesy of Brian Boucheron)

Start slow. Blend up some butter, make some slow-cooker yogurt. Try your hand at cottage cheese and some labneh. If you enjoy yourself and the food you are making, then really explore the possibilities. Find recipes, try new things. Herbed butter, kefirs, flavored cheeses. Everything happens one step at a time, and you never know what is possible until you give it a try.

Fibers: From Sheep to Sweater

22

The children's book *Farmer Brown Shears His Sheep: A Yarn About Wool* follows a bewildered flock of sheep as they chase the once-trustworthy Farmer Brown, who has taken their wool from them. Shivering and confused, they trek all over the countryside, stopping several times, only to see their warm wool become stranger and stranger.

In this charming story, everyone has a part to play. Rather than doing it all himself, Farmer Brown allows his friends to do what they do best. One cleans and cards the fleece; one spins the roving into yarn; and one dyes it all in bright, springtime colors. When Farmer Brown realizes that the sheep are shivering and quite concerned, he heads right back home to knit them all sweaters from their own—now very different—wool.

At the end of the book, the sheep and farmer are happily clad in hand-knit sweaters and looking forward to when they can start the process over. Readers end the book with a feeling as warm and fuzzy as the sweaters, but we also leave with a bit of an understanding of what it takes to turn your sheep's—or goat's or rabbit's—winter coat into your own coat or mittens: a lot of work.

An animal's fleece is not ready-made to spin. Obviously, they have been out in the elements for the last year. Diet and care will also influence the quality of the fiber that you end up with, as will the kind of animal you are raising. Check back to Chapter 14 for a rundown on raising sheep and goats and Chapter 13 for rabbits.

Additionally, many plants have been harvested as fiber throughout history, and still today, bamboo, flax, and (surprise, surprise) cotton are beloved sources of fabric. It takes space in your backyard garden and quite a bit of effort to turn plants into spinnable fiber. Animals are an easier start for the backyard farmer. Still, it might be a fun venture if you are already growing something like flax for its seeds or another beneficial aspect.

No matter the source, seeing your hard work culminate into wearable clothing or a marketable product is a unique joy and one worth the effort that goes into it.

Cultivating and Choosing Good Fleece

When assessing your flock's potential, you can safely assume that you will only be able to use about half of what you shear from your animals. If you are looking at plant production for spinning, you will need a significant portion allotted for growing or you will put all that work in for a single scarf! This chapter focuses mainly on animal fibers, but most of the dyeing and spinning principles will apply to plant products as well.

Handspun fibers are a treat to work with and create beautiful, valuable products.
(Photo courtesy of Theresa Armit)

For good, useable fiber, there are a number of factors. The number one, obvious thing to watch for is cleanliness. However, don't judge a sheep by its outer coat! Animals exposed to the elements—such as alpaca, who love to roll around in the dust—will likely have some unappealing qualities at first glance.

You will not use all the wool or fur, so know where to look and know that you will most likely be giving the fleece a good wash before working with it.

When selecting a fleece or monitoring an animal for their quality growth, you are looking for the general size (measures both length in inches and width in *microns*), color, elasticity, and character of the fiber. These elements will all factor in to the way it spins up, and there are benefits to many different types, depending on what you are looking for.

Longer fibers will be easier for a beginner to spin, while shorter fibers will produce a more delicate yarn that is difficult for a new spinner to work with. Each has its benefits and drawbacks.

A good fleece will be relatively clean, without excessive amounts of short hairs and unusable fiber.
(Photo courtesy of Steven Depolo)

Similarly, the shade of the fleece can affect how you are able to use the end-product. If you plan to dye the fleece, brighter whites will return bright colors. On the other hand, darker colors might be desirable to just spin without dyeing at all. There is definitely a growing trend toward naturally colored fleeces that don't require dyeing.

Another readily visible factor when analyzing a coat or a fleece is the amount of *crimp*, or character, that it has. This refers to the amount of wave and curl that you see in the fibers and will have an effect on the way that the fleece spins up. The crimp will also affect how the final product looks and feels.

Handspinners sometimes prefer to use fleece in its natural color, but only light colors will take dye. So darker fleece can be more desirable or less depending on what you want to do.
(Photo courtesy of Emma Jane Hogbin)

Preparing a Fleece for Spinning

Shearing time produces a single fleece from each animal. Laid out flat, you can see all the fiber that the animal has to offer, for better or for worse.

The process of removing undesirables is called *skirting*. The edges where the belly, neck, and legs were will likely need to be removed, as will any short, straight hairs that interrupt those good-quality fibers we talked about. Skirting your fleece will make a big difference in how your fleece turns out. There's a huge difference between soft, consistent, downy products and scratchy, difficult ones.

HAND-HARVESTED FIBERS VS. SHEARING

At one time, and still in some parts of the world, sheep and other fiber animals are brought indoors during their shedding season. Restrained, their wool is combed and collected as it is shed, separating the soft underside from the coarse outer hairs as it's gathered. The difference in the fibers is so marked, in fact, that the fabrics discovered around the time that shears had been invented were entirely different from the fabrics earlier on. Researchers speculated that coarser horse hairs or other fibers had been incorporated, until they realized that people were simply shearing and using an entire fleece rather than separating the soft wool from the coarse hairs.

Washing

If your fleece is coming from a sheep or goat, it will likely have a sheen to it and some mats or clumps. Dirt and whatnot are to be expected. This is, after all, an animal's coat. However, wool-bearing animals produce a substance in their sweat called *lanolin*—you might have heard of it for its other uses. If so, you know just how thick and greasy the stuff is.

Now, it is sometimes preferred to actually use the wool "in the grease," with the lanolin, or most of it, intact. This makes a very water-resistant wool, and some crafters might prefer it. If that is the case, you can clean the fleece of foreign matter by simply soaking it in cold water and removing it to dry. However, for easier spinning and use, a lighter texture, and the ability to dye your fleece, removing the lanolin is a must. And that is a bit more involved. It's water-repellent, remember?

Carefully dipping the fleece into hot water, sometimes with a bit of gentle detergent, will eventually strip the wool of all oils and residue. Keep a close eye, though. If the water cools too much with the fleece in it, the lanolin could bind to it again. If you agitate it, it will begin to *felt,* or mat together, and be more difficult to spin into yarn.

After soaking a few minutes, remove the fleece, empty the water, and refill the tub or basin. Literally rinse and repeat. After a couple of soaks, the water will run clear when you remove the fleece, and it will be clean. A delicates laundry bag may help you dip the fleece in and out without too much handling.

Take caution where you wash your wool as well. Too much lanolin in sewage pipes could cause problems. An outdoor area or basins that you can empty outside are a safer bet.

On the other hand, if your fiber comes from an alpaca, rabbit, or another animal (or plant!) that doesn't produce lanolin, you could just give it a good shake or a quick soak in cool water to get the dust out.

When washed, you certainly could go straight to the spindle or wheel. Often, though, you will want a more evenly distributed material to work with. That is where carding and combing comes in.

Carding and Combing

Both *carding* and *combing*—usually you will need only one or the other—can be done on as small a scale as with pet brushes or as large as big, industrial machines. You'll want to choose your tools based on the size and frequency of your projects, as well as your budget and space allowance. As with all things backyard farming, there are always bigger and more expensive machines available, but they aren't always worth the cost.

DEFINITION

Carding is a process that gets fibers ready for processing. Like brushing your hair, it gets out the bulk of the tangles, but because the fibers aren't fully lined up, it tends to leave the fleece more fuzzy in texture. **Combing** prepares the carded fibers for spinning. It is more labor intensive than carding because it aligns more of the fibers in the same direction, giving them a smoother, shinier texture.

I will stop here and say that the fiber arts, as they are so lovingly called, can be as thrifty or as pricey as you want to make it. From start to finish, every step of the way can be DIY and inexpensive or outsourced, purchased, or otherwise top-of-the-line. Neither way is right or wrong. Just don't write off spinning because you think it is going to be an expensive hobby.

With that said, you truly can card your fleece with just a pair of pet brushes, or comb it with pet combs. Both combing and carding should be a smooth process. Agitating the wool by grating bristles against each other or tugging at mats will only make a mess. The principle behind the process is to gently smooth the fiber into a fluffy, even consistency. Even plant fibers go through a sort of combing and carding to achieve a good consistency. This can be achieved by "flicking" a mini carder (or pet brush) lightly through some of your fleece as it lays on your lap, or gently passing mini combs (or pet combs) through it crossways. Slightly larger-scale combing and carding will cost a bit more and involve some more technique as you avoid meeting bristles, but it would be more efficient for a larger project.

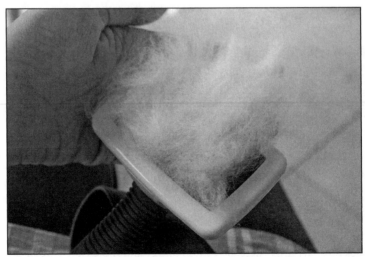

Carding and combing changes the way fiber will spin. This can be done even on the smallest of scales with a pair of pet brushes or combs.
(Photo courtesy of Brannan Sirratt)

You will not likely place an industrial-sized carder in your backyard farm, but a drum carder might be a worthwhile investment if you plan to do this for any extended amount of time. Using a drum carder involves picking the fleece to open and fluff it up a bit, then cranking the handle of the carder and letting the machine work its magic. It may take a couple of times through the carder, but it is much less work-intensive than hand carding. Of course, you will pay for that in the initial investment. But it may be worth it for you to save the time and energy.

Carded or combed wool can be referred to as *rolags, batt, roving,* and *top,* depending on the technique used and the end result.

Dyeing Your Fiber

Unspun fiber is a treat to dye and offers quite a bit of creativity. As with washing, you will need to take great care to prevent felting if you plan to spin it later. Agitation is your enemy, as is a sudden change in temperature. Even spun yarn can and will felt up on you, so don't think that will save you in the dyeing process. It can be just as fun and useful to dye spun yarn as unspun fiber, but as long as you are using animal fibers, you have to be aware of the felting possibility.

Earlier, I mentioned the shade of white and colors of a coat. The brighter the shade of white, the better it will hold color, especially bright color. Depigmentation lightens darker shades to make dyeing easier. This usually involves harsh chemicals, so you might not want to delve into that in your personal venture.

Still, you will want to consider the shade of your roving alongside the color(s) you are planning to dye. A darker shade will dampen the color's tone. The texture and density of the roving will also have an effect on the brightness and way the dye is taken.

Not only that, but the type of dye you use will also have different effects. A basic dye recipe simply involves soaking your roving in a vat of sugarless Kool-Aid until the water is clear and the roving is bright. However, you are unlikely to get a gentle sage-green from this method, with color names like "Sharkleberry Fin" and "Kickin' Kiwi-Lime." If colors are steam-set well, they will be permanent.

Truly natural dyes—like the kind you find in your garden—will give you more muted results. Acid dyes will afford the most control over the final color.

Color aside, many factors can affect dye choice, and you may use a different type at different times. Dig in, experiment, and see what works for you.

OVER THE GARDEN FENCE

Your garden is not only a source of food for your family. It can also be a source of dye for your roving! Consider red cabbage, berries, dandelion flowers, marigolds, onion skins …. Anne Bliss, author of *North American Dye Plants,* offers this advice: "Yellow-flowering plants most often produce yellow-cast dye, as do most white bloomed species. However, some white-flowering plants yield tans. Plants with purple blossoms usually give tans, golds, and greens. Red berries generally produce yellow or golds."

Before dyeing your wool, you might want to *premordant* the fleece—that is, treat the fibers with a solution that will prepare it to hold color longer and more vibrantly. Alum powder and cream of tartar, iron in various forms, and copper sulfate are examples of mordants. You might also soak the dyed fleece in vinegar or baking soda to add another layer of color enhancement and setting. Baking soda will bring out blues; vinegar enhances reds.

The dyeing process carries a similar caution to washing: do not agitate. After your fleece is in the water and dye bath, leave it alone. After the color has been sufficiently absorbed, remove the fleece and carefully rinse it, starting with temperatures similar to the bath and slowly cooling it until you have cool, clear water coming off of it.

Drying is equally precarious, as too much handling and squeezing can affect the shape of the wool or even begin to felt it. It's easy to dry hanging loosely over a hanger or on a screen so the fibers aren't disturbed. Just don't twist it or wring out the water as you might with a towel because that harsh action will ruin it.

With that said, a dye bath can have many faces. You will want to find out what exact amounts of water and dye will work for your specific project, and then get creative with the process!

Hands-Off Dyeing

If you are not worried about monitoring the wool to get a light shade, dyeing in the oven or in a slow cooker is the way to go. With vinegar in the bottom, coils of roving layered with sprinkles of dye, and then more vinegar and lots of water carefully poured in, your slow cooker will simmer for hours while your fleece soaks in gorgeous colors. No need to watch or stir or worry about the kids getting into it.

OVER THE GARDEN FENCE

You probably won't need a separate slow cooker if you are only using teas and natural dyes to color your roving. However, if you are dyeing with synthetic or purchased dyes, you will want a slow cooker dedicated to dyeing that isn't used for foods. The same would go for regular pots on the stove using the stovetop methods. I use an old pot with a broken handle that I'll never cook with again.

Similarly, you can soak your roving in some water and vinegar, lay it in a casserole dish, cover with dye dissolved in water, and then bake it at 350°F for about an hour. A quicker version is obtained by putting the fleece in a microwave-safe dish with water and dye, then nuking it for five to six minutes or so. Each of these methods will get you the full potential of the dye. By the end, the water should be clear and the wool should have soaked up the entirety of the dye.

Stovetop or Vat Dyeing

If you would like more control of color or do not want to mess with ovens or buying a slow cooker just for dyeing, you might enjoy vat dyeing or stovetop dyeing. Dyeing in a vat simply means immersing presoaked fibers in dye and hot water. The goal is to keep the water hot, not necessarily simmering, so this might be a good outdoors project on a summer day. For stovetop dyeing, you want it to simmer consistently for 30 minutes to an hour or less, depending on the color results you are aiming for.

Hand Painting with Dye

For even more control in your final product and the opportunity for even more creativity, consider hand panting the colors. One method involves laying strips of soaked roving onto plastic wrap, literally painting your dye mixes on in whatever pattern or blend you'd like, then rolling it up in the wrap and steaming it in a canner.

You will never be able to fully control the colors and patterns you develop when dyeing fibers. I think that may be a big part of its appeal.

Hand painting yarn is a lot of fun and gives you more control over the end product. Wrap in plastic wrap and steam to set the dye.
(Photo courtesy of Jen Dickert at SupernaturalNutrition.net)

Spinning Yarn

As long as humans have used tools, they have spun yarn. Spindles have been discovered that pre-date the wheel. Maybe the spindle even led to the wheel! At its most basic, spinning fiber into yarn can be achieved by hand. Tools will almost always make a job easier, but start by handspinning a bit of fiber to get a feel for it. From there, you can slowly move up the proverbial ladder until you find your spinning niche.

Spindles and Wheels

Drop spindles are so simple that you can make your own with a CD and a dowel. There are variations in hand spindles, but the drop spindle itself will hang suspended, spinning in the air, as you use both hands to draft and control the fibers and yarn. You can find them inexpensively with some roving online. This makes them a good choice for a beginning spinner to get the feel for drafting. Yarn shops will often hold beginning spinning classes that cover using drop spindles.

For bigger or more frequent projects, or simply because you are hooked and want to take the next step, a wheel is in your future. Surprisingly enough, you can find tutorials to make a wheel, so the DIY aspect is not limited to spindles. If you are purchasing a wheel, you will need to take many factors into consideration and take your time deciding. WorldinaSpin.com has a lot of tutorials for specific fiber art techniques.

Try them out at the store. Use a friend's. Rent one. Borrow one or two. Try as many as you can before purchasing to make sure you are comfortable with the one you settle on. Not only are there budgets to consider, but size, shape, whether you are left-handed or right-handed, and the type of yarn you expect to spin on it are all important to consider as well.

Try out several wheels before buying one so that you get a feel for what will work for you.
(Photo courtesy of Chris McLaughlin)

Carded, medium to long fibers are the easiest to begin on. The reason for this is simple: you will be doing a lot of multitasking to spin your yarn, and the last thing you need to worry about is clumpy roving. Fluffy rolags of wool are good starters, because they will draft well, which we'll look at next.

Spinning Terms and Techniques

Drafting is the first step to spinning. Pulling some of the fibers out from the rest is called *drafting;* giving it twist is called *spinning*. As you draft and spin, the wool stretches and moves and turns into yarn as you know it. Whether you spin gently on a drop spindle or rhythmically with the treadle and wheel, the twist is the difference between roving and yarn. How it is drafted and spun will determine what kind of yarn you end up with:

- ✣ **Novelty yarn.** When you're comfortable spinning, the itch to create might strike you. Subtle changes such as the varying thickness of the yarn, to dramatic novelty techniques such as adding in ribbon or glitter, will keep you busy indefinitely. Depending on the drama, novelty yarns could sell for more, but they could also be sold less frequently, as they will not appeal to so wide an audience. But it never hurts to play!

- ✣ **Plied yarn.** Spinning can and certainly often does end at just the first spin. However, *plying* it—putting a reverse spin on two or more strands to join them together into a stronger strand—can be quite valuable. Plied yarn is stronger, more even, and more forgiving in use. A spinning wheel makes this process relatively painless, as it will do the reverse-spin for you. Because plied yarn makes much more yarn out of your fleece, it is likely worth the bit of extra finishing time it will take to ply.

- ✣ **Woolen yarn.** Shorter fibers carded into fluffy rolags will spin well into woolen yarn. Rather than spinning from a dense draft, you will modify the technique to allow for some air to be trapped in the fibers. This makes woolen yarn light, airy, and lovely for soft garments.

- ✣ **Worsted yarn.** This is the most common technique and the best for beginners. To spin worsted yarn, you must use longer fibers that have been combed or carded—the mini combs and flick carder discussed earlier in this chapter work very well. You will then draft the fiber before the twist begins, flattening it as you draft. Worsted yarn makes durable, warm fabrics.

When your hank of yarn is finished, you can set the spin by soaking the yarn in warm water, then squeezing out the water and hanging it to dry. If the yarn hangs freely and doesn't curl up or into itself, it is set. If not, try soaking again and hanging a weight on it when it is drying.

Congratulations! Your yarn can now be rolled into skeins and sold or used for crafting.

You cannot compare handspun yarn with commercial yarn, especially commercial synthetics. The quality is simply on a different level.
(Photo courtesy of Emma Jane Hogbin)

Marketing Your Fiber or Yarn

You probably don't want to knit sweaters for your livestock, but I think you can learn something from Farmer Brown. Sometimes, it's wise to let others do what they do best. If you know a spinner and you really don't have the time or interest, see what you can work out with them—especially if they enjoy it as a hobby. They might purchase whole fleeces straight from you, saving you the time and work that goes into preparing and spinning.

Perhaps you like to spin, but aren't interested in the preparation. It might be worth it to you to hire some teens or fiber enthusiasts to prepare it for you, and then you can spin and sell the yarn. Here, you will need to take into consideration how much you can sell the yarn or finished products for as compared to what you will spend on preparation. Your time could be worth the difference.

Knitters, crotcheters, weavers, and felters all love to work with handspun yarn. Consider selling your wares at a local farmer's market, to a yarn shop, online, or to friends. If you enjoy crafting with yarn, you might find that the finished project holds more selling value than the materials.

Ask around. Try different avenues. But most of all, attain and maintain a high quality of workmanship, and you'll likely find many happy buyers.

Home Brewing Cider and Wine   ∽23

My uncle lives on a harbor in Florida and his backyard space is smaller than many of the driveways in my city area. But his beehives provide him honey from which he is able to brew homemade mead in both dry and sweet recipes.

It was summer, and my family was traveling to Florida for a visit. The trip fell on my twenty-first birthday and I remember toasting the occasion with homemade brew. As I held my glass and heard words of blessing and celebration spoken over me, I felt so special and grown up. I was being toasted by my family with something that had been carefully created with care, time, and love.

Home brewing techniques rely on fermentation, which is a way of preserving. It preserves fruit by converting the sugars and carbohydrates into alcohol. All of the basic recipes follow the same general template: fruit juice is mixed with sugar and yeast, and allowed to ferment. Then the wine ages and is strained, bottled, and enjoyed.

Brewing Homemade Cider

Hard cider, which is cider with alcoholic properties, should be made from a mix of apple varieties. Soft cider or sweet cider is the fresh-pressed juices of the apples; hard cider is what results after the fermentation. Cider is different from apple juice in that, among other things, it isn't filtered so apple particulates are still present.

It has been said that cider making is one of the best historic ways of preserving the food value of apples over the winter. While many apples will store well in cool cellars for weeks into winter, you can't store any that are bruised. So pressing them into cider is the perfect storage solution. It's definitely one great way to use extra apples from your home orchard while the harvest is good.

OVER THE GARDEN FENCE

If you don't have an overabundance of apples for yourself, you might be able to pick up windfalls from a neighbor's apple tree (ask first, of course!). Pick-your-own orchards are also good places for you to find locally grown apples at an inexpensive price.

Cider making is a seasonal pastime in many cultures. October is National Apple Month in the United States. In Britain, October 21 is "Apple Day." It's no wonder that so many communities come together for pressing cider in an old-fashioned gathering much like a barn-raising or quilting bee. Cider making can be time-consuming, but also worthwhile.

Pressing Your Own Cider

To press your own cider, you'll want to have at least 10 pounds of apples ready. One gallon of juice takes somewhere between 10 to 15 pounds of apples.

You will need a few items to make your own cider:

- Grinder or crusher
- Cider press
- Pressing cloths and filter cloths
- Collection bucket or vat (should be stainless steel or enamel glazed to prevent unwanted interactions with the apple juices)
- Funnel and tubing
- Storage containers

Apples are best pressed into cider when they've been able to mature or "sweat" as it is often called. You'll know the apple is ready to press when the apple has a little bit of give to it when the skin is pressed. Separate different types of apples when you press them because each variety of apple has its own taste and flavor. They even have different chemical make-ups.

Tannin is what gives some apples a tarter, dry taste that leaves a bitterness after eating. Mixing high-tannin apples with various sweet and acidic apple varieties will change the taste and complexity of the finished cider.

Professional cider makers (and obsessed hobbyists) press each variety of apple into individual containers and then mix them together in special (often top-secret) blends. This gives them more precise control over the final outcome instead of just tossing all the apples together and pressing them at the same time.

While any apple can be pressed for cider, here are some classic varieties known for contributing to excellent ciders:

- Baldwin—Medium-acid and sweet flavor.
- Fair Maid of Devon—High volume of juice.
- Golden Delicious—Sweet and light flavor.
- Jonathan—Balanced sweet and sharp flavor.

⁜ Kingston Black—One of the best cider apples, great bitter-sharp flavor.

⁜ McIntosh—Tart and tender.

⁜ Royal Russet—Sweet yellow flesh with hardy flavor.

⁜ Sweet Coppin—Sweet and bland, a good mixing juice.

Look for a good mix of sweetness and acidity when you mix your ciders. If you use only sweet apples, your cider will be bland. If the cider is too sharp, it may be too strong when the cider is fermented.

After you've selected and gathered your apples, the fun (some people call it work) begins. Wash your apples thoroughly in clean water and then add them to the grinder whole, core and every-thing. The grinder will crush the apples into a sort of mush called *pomace*. The pomace should be pressed immediately after grinding.

After the apples run through the grinder, the pomace will drop into the bucket. Note the lining in the bucket, which allows you to switch it straight into the presser.
(Photo courtesy of Ben Garney)

Move the bucket with the pomace to the cider press. Usually this is a screw press or ratchet press in simple home setups. Close the press bag over the pomace and slowly press the pomace in the cider press. Continue tightening the press down as the juice flows so you can squeeze out all the juice. Use a nonreactive bucket to capture the juice. Stainless-steel, plastic, or enamel-coated containers are good choices.

Now press the apple pomace and capture the juice in a bucket, tub, or nonreactive container.
(Photo courtesy of R. M. Siegel)

Pour the juice through a strainer cloth like cheesecloth or nylon mesh. This removes any impurities that may have slipped through the press and any pieces of pomace.

Filter the juice through a strainer to catch any leftover impurities, if needed, and begin filling your jars.
(Photo courtesy of R. M. Siegel)

After you've strained the cider, you can fill your jars. Try clear glass or plastic jugs that are completely clean and sanitized.

Fill the glass or plastic jars with your finished soft cider.
(Photo courtesy of Ben Garney)

If you choose to store your cider at this stage, before fermentation as soft cider, you can store it in the fridge for a couple weeks, and in the freezer for much longer. When freezing your cider, allow 2 to 3 inches of headroom at the top of the jug to allow for expansion during freezing.

Fermenting Your Apple Cider

To make hard cider, you'll need to start with apple cider of some kind—either store-bought or home-pressed apple cider that you've made yourself. If you choose to purchase apple cider for fermentation, be sure it doesn't have chemical preservatives because that will inhibit the growth of the yeast. If you press your own cider, start the fermentation process right away.

I discuss a simple fermentation method, although there are several various methods you can use if you decide to get fancier later. Be sure to use careful sanitary measures to avoid contamination by wild yeasts and bacteria (see the following sidebar).

Bring the cider to a simmer in a nonreactive brew pot, to allow it to heat up for about 30 to 45 minutes. Add in some sugar or honey if you want to increase the fermentable material. Do not boil the apple cider or the flavor will change. This simmering stage is not required, but will help the yeast grow faster.

STERILIZING YOUR EQUIPMENT

Sterilizing something is different than simply cleaning it. When you clean something, you are removing visible dirt and debris, and for many household chores that is enough. For sensitive projects like cider making, milking, and canning, sterilizing the jars and other equipment is an important first step that kills microscopic molds, bacteria, and other contaminants. These can not only affect the flavor of your products, but make you seriously ill as well. You can sterilize your glass jars or other equipment in a couple different ways. If your dishwasher has one, choose the "sanitize" setting to heat the equipment to a level that will kill contaminants. Alternatively, you can use your water-bath canner to boil the items for 10 minutes.

Pour the cider to be fermented into a large drum, like a 5-gallon plastic bucket with spigot and air lock. When the cider has cooled to nearly room temperature, add the yeast. For 5 gallons of cider, you'd want two packets of yeast. Five quarts of cider would be given about 2 teaspoons of yeast. It's not recommended to use bread yeast, but rather one of the wine yeasts you can purchase from a brewing supply store.

The first, or primary, fermentation can produce a lot of bubbling and brewing in the process. This cider-maker has a drainage tube to allow the escape of gasses and foam.
(Photo courtesy of R. M. Siegel)

Let the cider ferment in a cool place between 60°F and 68°F for two to three days. You should notice some bubbling and frothing in the cider—this is the carbon dioxide being released through the yeast fermentation process. Cover the air lock with a damp cloth to allow this byproduct to escape and remove the impurities. After things calm down the second or third day, replace lost volume with room temperature water and secure the air lock.

Now let the fermentation continue for another few weeks. It should take about six weeks or more and you'll notice that the fermenting activity will have slowed a great deal. There will be yeast debris at the bottom of the jug that you wouldn't want to use.

At this point you can bottle the cider now after one fermentation cycle, or you can use a process called *racking off* to go through another cycle to make the cider more clear. If you choose to further refine the cider, continue with the following steps.

Racking Off the Cider

To clarify the cider, or take it from murky and opaque to a more crystalline appearance, you'll want to further ferment the cider. Siphon it to another jug or bucket to allow the cider to ferment again and separate the liquid from the *apple lees* (dirty debris that's gathered at the bottom of the bucket) and yeast deposit that has piled up.

After you've moved the cider to a second fermenting bucket, top it off with cold water to replace what you've lost. If you want a sweeter taste to your finished hard cider, add sugar or honey to the water before mixing it in. Allow the cider to ferment again in a cool location for several weeks. The longer you let it ferment, the clearer it will become as the suspended yeast clears away. To bottle the cider, siphon off the liquid into a sanitized glass or plastic jar.

MAKING SPICED CIDER

Simmer 1 quart of cider with mulling spices: whole cloves, cinnamon stick, cardamom seeds, and a pinch of nutmeg. Add a peppermint leaf or bay leaf if desired. Allow spices to steep in the cider for 20 minutes or more. Strain out the spices, bottle the flavored cider in a large sterilized jar, and store immediately in the fridge. Use within a week.

Brewing Homemade Wine

In this section I introduce the basic steps of wine making. I recommend *First Steps in Winemaking* by C. J. J. Berry for more details if you get hooked on making wine.

Before investing in expensive equipment, start simply, and try out a round of wine making. These more commonly found items are a good place to start:

- ✢ Saucepan (nonreactive)
- ✢ 10-quart bucket
- ✢ Siphon tubing
- ✢ 5-pint jug with cork and fermentation lock

- ✢ Funnel
- ✢ Strainer
- ✢ Yeast
- ✢ Yeast feeder
- ✢ Citric acid
- ✢ Sterilized wine bottles

You can usually find inexpensive starter kits at wine making shops, feed stores, or country supply shops. If you use pots, buckets, and pans from home, be sure you use nonreactive materials like plastic, stainless steel, or enamel coated. Glass is the historical material of choice and doesn't have the ability to alter the flavor of your wine.

While you can buy fruit juice concentrate from purchased kits to make wine, the result will be far inferior compared to making your own juice. Grapes are the traditional fruit for making wine, of course, but other fruits will work as well, such as elderberries, peaches, and apples.

You can crush your grapes in a cider press, with a mortar and pestle, or the iconic foot stomping seen in movies and magazines. Collect the juices and add citric acid or lemon juice (either bottled or fresh), sweetener or yeast food, and high-quality yeast. Your wine making kit should have directions for the exact amounts of each. Homemade recipes call for a variety of sweeteners depending on your preference—for example, sugar, honey, corn syrup, or fruit juices.

I recommend using a yeast designed for wine making if possible. Lemon juice or citric acid will increase the acidity of the wine to help it ferment properly. This original combination of juice and yeast mixture is called the *must.* Add everything to a clean container and stir it thoroughly. Leave it at room temperature for one to three hours until it begins to bubble and foam.

Put the lid on the container and add your fermentation lock to the bucket. Allow the must to ferment for a few days while it actively bubbles and foams. You may see a layer of particulates form at the top of the bucket, carried by the foam. Sediment, called *lees,* will also pile up at the bottom of the bucket. You'll separate this out during the racking process.

Siphon off the liquid from the first bucket into a clean and sanitized second container. This second container should be set up just like the first with an air lock or fermentation lock. Do not siphon off any of the lees or solid particulates. Allow this second container to ferment further until the liquid turns from cloudy to clear.

THORNY MATTERS

If the wine still seems to have some fermentation activity, you'll want to do a third racking. Too much fermentation after bottling could actually burst bottles and pose a potential danger. A third racking can help prevent this problem.

It can take a few weeks or it can take a few months, but you'll know the wine has finished fermenting when the liquid turns clear. At this point it's ready to bottle and store. Be sure the bottles are completely cleaned and sanitized before bottling. You can bottle in bulk in a large container, which keeps the wine more stable. Or you can bottle in individual bottles and corks. Either way, supplies are available through a winery supply store. You should always purchase new corks for your wine to avoid contaminating the new wine you've created.

Store your wine in a cool, dark location to preserve the best flavor. A root cellar is perfect, and there's a reason why wine cellars, and not wine attics, were the norm.

Goat's Milk Soap Making

Soap making is a centuries-old art that, at its core, has changed little over time. Ancient Babylonians would combine oils with ash to create soap. Today, oils and lye are used. Either way, it is the reaction between an alkaline substance and oil that begins *saponification*—literally, "the making of soap."

Creating a chemical reaction in your kitchen may sound intimidating, but it isn't any more ambitious than frying something in hot oil. You should use precautions to keep small children safe, and wear protective gloves and eyewear just in case, but cooking is cooking—only this time, you are making a bar of luscious soap instead of a platter of chicken-fried steak.

Health Benefits Infused in Soap

What would the cumulative effects look like for a wave of people to switch to homemade soaps? For an individual, this will help you avoid potential toxins and actually infuse a number of benefits into your skin, and therefore your body.

A plain soap made from coconut oil, olive oil, or almond oil can be refreshing. A goat's milk soap adds another level of moisture and can also include vitamins, including vitamin A, which repairs and maintains your skin's health; alpha hydroxy acids, which wash away dead skin cells naturally; and even minerals such as selenium, which actually work to fight off damage and even skin cancer. Herbs and essential oils stack even more benefits onto the list. The choice between carcinogens and skin damage versus immune builders and skin repair should be a no-brainer.

For most people, it isn't the knowledge but the accessibility that is the roadblock. Goat's milk soap might not be readily available or as seemingly affordable as their normal brand, even though natural products tend to last longer. The thought of making your own soap seems rather daunting, so it isn't usually thought of as an option. But soap making is easier than you think, and once you try it you'll never want anything else for your family!

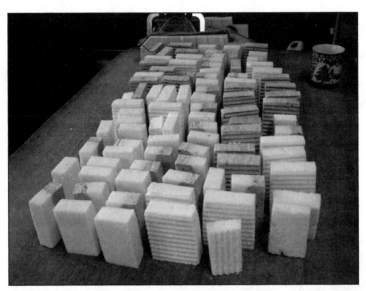

Soap can be made in large batches, so you don't have to make it as frequently (unless you enjoy doing so, as many people do).
(Photo courtesy of Theresa Armit)

Homemade soap can be filled with oatmeal, herbs, flowers, essential oils, and of course goat's milk to maximize the benefits to your skin.
(Photo courtesy of Tori Deppling Roudy)

Why Make Your Own Soap?

Unfortunately, when it comes to commercial soap making, sturdiness is more valuable and tallow more readily attainable than more exotic oils. So artificial detergents, moisturizers, and lathering agents are likely to be added. And it is not usually just one. If you include a harsh detergent, you have to include a moisturizer. If you include a moisturizer, you need to make sure it lathers well. To make it lather well, you might need to make sure it breaks down hard water. On and on you go until the ingredients list is long and filled with what might as well be another language.

Those ingredients are more than just words. Your skin is the largest organ of your body, covering upward of 8 to 10 square feet and comprising as much as 15 percent of your body weight. Now, think about what you put on your skin. What's more, is there anything you apply to your skin more thoroughly than soap? You wouldn't rub pesticides onto your liver. Or seep preservatives into your lungs. Your skin should get the same level of care.

To top it all off, imagine all of those toxins rinsing right off of your body, down the drain, and ultimately back into your drinking water and the earth in general. Nothing that we do exists in a bubble. The cumulative effects of millions of people rinsing carcinogens, pesticides, formaldehyde, and who knows what else down their drains will come back to haunt us.

Making Goat's Milk Soap

Goat's milk soap is a great way to use up the excess milk from your backyard farm, but there are a few extra things you'll need on hand to make it. You'll need a scale to measure your ingredients precisely, a large stainless-steel pot that will not corrode from the lye, an immersion blender, a thermometer, and a couple hard plastic spoons.

THORNY MATTERS

Label your pots, blender, and any other equipment that you'll be using for soap making with a permanent marker so it doesn't get mixed up with any dishes or other utensils used for preparing food.

Other items that do not need to be for soap only are gloves and protective eyewear, as well as a mixing bowl for your herb and oil additions. Finally, you will need something to mold the soap in and cover it as it sets. A cardboard box will work, though you can buy molds at craft stores.

Follow a recipe closely at first, and over time you will become familiar enough with the ratios of oil combinations to lye, on top of the addition of milk and any other ingredients you would like to add, to create your own concoction.

Measuring your ingredients perfectly is vital to good soap. With imbalanced ratios of oil and lye, the soap could be too harsh or too heavy. A good kitchen scale makes all the difference—make sure you know how to use it before you get started. You need a measuring container on the scale, with the scale zeroed out before you begin to pour. Go very slowly and make sure you don't get too much. When you are mixing a bowl of several oils together, there isn't really a way you can get back what you've poured in!

Measure carefully so you're sure to stir in the right amounts.
(Photo courtesy of Steve Bozak)

Again, your recipe will tell you how much to use at first. But as you make your own, you will find that they are all based on ratios, such as part oil that makes a hard soap, part that will be moisturizing, and part that will be lathering. Calculating the lye percentage is a bit different. I won't even try to list here the math that goes into it. Suffice it to say, if you hope to make your own recipe, there are numerous calculators online that can run the numbers for you and tell you exactly what you need. TheSage.com is one such resource.

For goat's milk soap in particular, not just any recipe will do. It is true that the goat's milk replaces water in a recipe, but not just any cooking process will work. Some soap can be made with a high heat process that moves and cures a little faster than a cold process. The problem with that is that goat's milk will curdle at too high of a temperature, and that isn't something you really want to slather on in the shower. Using goat's milk will change the whole soap making process to prevent problems.

So to make goat's milk soap, you have to keep the oils cooler than 100°F. Saponification will happen when heated, though, so you do need a thermometer to help you gauge the temperature control. Another step of protection for the milk is to first freeze it. The frozen milk will be mixed with lye before combining with the oils, and lye is actually caustic enough to scald it. By freezing the milk, you can break it into chunks and then pour the lye over it (all while on the scale to ensure proper measurements) so that it slowly melts the milk and blends the two, rather than scorching it.

Lye can be harsh and burn the skin. This soap-maker is wearing gloves to protect himself from accidental burns. Making soap outdoors provides plenty of ventilation as well.

(Photo courtesy of Steve Bozak)

THORNY MATTERS

Lye releases strong heat, fumes, and gases when mixed with water. It will usually come in powder form, so be sure you have your skin and eyes protected, windows open and fans on, and children out of the room. Also, make sure the pot or pitcher you are mixing the lye solution in can handle the heat release without reacting. Use only heat-resistant plastic, stainless-steel, ceramic, or stoneware containers for blending the milk and lye. Vinegar will help counteract burns if lye touches skin.

Saponification

Saponification basically draws out the salt of the fat. Knowing this might make some of those confusing ingredients clearer, such as sodium tallowate as being the salt (soap) derivative of tallow (beef fat). When oils have been saponified, it creates soap and glycerin. In commercial soaps, glycerin is often removed. Most homemade soaps leave the glycerin in.

Saponified with sodium hydroxide (lye), soap will be hard. With potassium hydroxide (potash), soap will remain liquid.

Because it is simply fat that is needed to combine with an alkaline, both animal and vegetable oils can be used. Tallow and olive oil are common choices, though a few oils are usually combined together for their various benefits. Coconut oil, for example, will lather; olive oil will moisturize.

The so-called "cold" process of saponification … isn't. Heat is still involved, as well as two pots you don't mind losing to a soap-making addiction. One will be for the lye and milk, and the other will be for the oils. Before you put anything on the stove, make sure your molds are ready to fill with soap as soon as it is ready. Small boxes and pans can work, as well as muffin tins. Line them with wax paper or grease them with a cheap cooking spray, though, so the soap doesn't stick when you try to pop it out of the mold.

Gently heat the measured oils in the larger of the two pots, watching the temperature carefully so that you don't make it too hot for the milk. Heating the oil will melt any butters or solids that you used. As soon as they are all melted together, or close, as is the case with cocoa and Shea butters, turn off the heat and set the pot in the sink.

If the oil is not hotter than 100°F, slowly pour the milk and lye mixture into the oil. It is important to do it in that order, as pouring the oil into the lye could cause a different and potentially danger-ous reaction. Stir carefully as you pour. Stirring is the catalyst here, and at one time, it would take an hour of stirring before getting it where you wanted. No wonder soap was a luxury! You can fill the empty lye container with vinegar and water so that the container will start to clean a bit while you stir your lye and oil mixture to help save time on the cleanup process.

Today, we have immersion blenders and stick blenders to get the job done for us. Stir it around in your mixture, then turn it on and blend for a bit. Go back and forth between stirring with and without the motor running. The blender is your very good friend, and you want it to last for many soap batches to come!

Thanks to the blender, you should start to see it thicken very soon. The goal here is something called trace, which is roughly the thickness of pudding. At this point, your soap will not come undone, so to speak. It will not separate back into lye and liquids. It is so named because if you drizzle some of the mixture back into the pot, it will leave a traced marking across the top where it drizzled.

When it is well mixed and nearing or at trace, it's time to add in any other ingredients you had planned to use: essential oils, dyes, herbs, flowers, what have you. You need to work somewhat quickly. After it reaches trace, it will begin to set up, and you don't want a pan-sized bar of soap. Pour the soap into the mold, and then wait.

Let it sit in the mold uncovered. Some instructions will have you cover or insulate it, but goat's milk produces heat, making covering unnecessary. In roughly a day or two, you will be able to remove the soap from the mold and cut it into bars. Wait until it is no longer smooshy when the knife hits it, but not so long that it is too brittle and difficult to slice.

This homemade mold holds a large batch of soap. Don't feel like you have to buy everything.
(Photo courtesy of Steve Bozak)

When the soap has slightly hardened, you can cut the bars. This large slab will become several bars of soap.
(Photo courtesy of Steve Bozak)

Make sure to wear gloves again, as the lye could still irritate your skin. Cut bars should be left out to cure for several weeks before using. Now, the saponification process will happen relatively quickly, in that 24- to 48-hour window. You might still have some irritation issues if you use it during that time, so it is safest to wait. On top of that, though, waiting allows any excess water that might be in the soap to evaporate. The more thoroughly this step happens, the harder and more resilient your bar of soap will become.

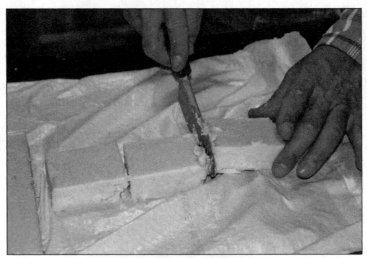

Now the soap can be cut into individual bars. Note that this soap-maker isn't wearing gloves. If the lye is still active in your soap, gloves are recommended.
(Photo courtesy of Steve Bozak)

ON A DIFFERENT SCALE

Rather than starting from scratch, you can purchase "melt and pour" soap at craft stores. It is just a basic soap to be grated, melted with goat's milk and any other goodies you want to include, then set until firm. No need to work with lye or wait for saponification!

Creative Soap Making

A simple bar of soap with a good blend of oils and rich, fresh goat's milk will be rich and healthful. Likely, there will be times when that is all you need. As with all art forms, however, this is not likely to be the case every time.

A sturdy workhorse soap will only need basic oils and does not necessarily need pretty dyes or extras like flowers and embellishments. But it may benefit from a refreshing peppermint scent and some exfoliating properties. A gift soap to pamper someone special might need lavish oils and a dash of flower petals.

Think about who you are giving the soap to and how each batch will be used. If you are making a large batch at once but have several needs, pour the trace soap into separate containers, quickly mix in what you need, and then set them in their respective molds.

Excess soap makes wonderful gifts or wares to be marketed.
(Photo courtesy of Tori Deppling Roudy)

Oil Combinations

The very first creative step in soap making is the oil blend. As I mentioned earlier, oils can be selected with a general knowledge of how they will react and what kind of soap they will make.

Palm kernel oil, coconut oil, tallow, and lard will all produce hard soaps. They each have varying benefits on top of that to help you choose between them. Lard and tallow might be more readily available to you if you raise meat animals, but the benefits of the others could be worth purchasing either way. You can mix hard oils, but just know that a soap that is too hard will be brittle.

Aside from durability, the next most important thing in a bar of soap is often considered the lather. That is why lathering properties are artificially added commercially. We don't tend to feel clean unless we have had a good lather. Coconut, soybean, and almond oil are among the oils that induce a rich lather naturally.

Finally, a really good soap will leave your skin moisturized. Goat's milk itself will actually contribute to this because it has such a great fat content to begin with. That means your soap is already ahead of the curve. Avocado, lard, and olive oil will accomplish this. Soap made entirely or largely from olive oil is called "Castile" soap. For extra indulgence, cocoa butter, Shea butter, mango butter, and other luxury ingredients can be added into your ratios.

Dyes and Embellishments

Particularly if you are marketing or gifting your soap, which will likely be the case if you make many batches, dyes and added ingredients can add to the visual appeal. Because you are doing this to use natural ingredients, preferably from your own backyard, and avoid synthetics, try remaining natural all the way through. Don't give up on your natural ingredients just for aesthetics.

Fortunately, you don't have to. Just about anything in your garden can be boiled down into a dye bath. (I talk more about plant-based dyes in Chapter 25.) You can add subtle and even deep color to your soap by including a homemade dye. Infuse your natural dye with one of your oils, or purchase a powdered version online or from a supply store. Natural dyes have a bit of variance to their results. It might fade a bit as it cures, but the general color should remain. If you are not happy with the color, you can always grate it down, melt it, and redye and set it.

Dyed or not, most will agree that fragrance is pretty much a must. Everyone loves a nice-smelling soap. Even though your soap will probably smell fresh and lovely anyway, essential oils are an easy way to infuse a customized scent, and making the combinations can be a lot of fun. Especially considering essential oils can provide more than just an aroma, but also health properties.

And that brings me to the next addition: herbs and flowers! Sprinkle some fresh herbs and flower petals into the soap for fragrance, health, and visual appeal. (Chapter 9 covers herbs in greater detail.) This is a great place to extend and exercise that knowledge, as well as use up some of that extra harvest.

THORNY MATTERS

Use caution when including essential oils and herbs, particularly when the soap will be used by children or pregnant women. For example, peppermint, jasmine, and rose, to name a few, are unsafe for use during pregnancy. Others, like cinnamon, orange, and wintergreen, might irritate the skin. The safest oils for pregnancy or sensitive skin are lavender, eucalyptus, chamomile, and lemon. One half ounce is usually enough for a pound of soap. You can experiment with more or less, but too much might seep out of the soap as it sets.

Let your creative side shine when you make soap. Did you dye with a blueberry dye bath? Add some dried blueberries into the soap. Maybe it is a citrus theme, so consider grating orange peel into it. Just be careful not to go overboard and cram all of your ideas into one bar. Adding a lot of solids can make an abrasive bar, which can be good for exfoliation or smoothing rough skin, but not suitable for sensitive skin or facial use. After you get a couple of batches under your belt and build your confidence and know-how, you can make all your ideas a reality.

Other Ways to Use Your Harvest ❧ 25

As long as people have cultivated the ground, they have worked hard. But that doesn't mean it was all work and no play. Hobbies emerge when there is downtime, and the hobbies of the farm are time-tested.

Imagine just 100 years ago, before TVs and blockbuster releases. What did the family do when they weren't working the fields, shelling peas, tending the animals, or cooking up meals? Put that way, it doesn't sound like much time was left at all. Still, people always have and always will look for some relaxation and enjoyment. The difference between then and now might just be that relaxation still meant productivity.

Do we need colorful clothing, flower arrangements, striking baskets, or homemade crayons? Maybe not. But we enjoy them, and you might just find that you enjoy making them. A key trait of a successful backyard farm is to not let anything go to waste. So here are just some of the things that can be done, and enjoyed, with the excess of your harvest.

Basketry and Weaving

Plants can be productive contributors to your garden or homestead long after the fruit is harvested and the flowers fall. Many of the leaves and stalks around your garden are built of strong fibers that can be turned into rope, baskets, trellises, and even yarn if you have the patience and curiosity.

At the end of the growing season, watch for the strongest leaves. If you can tug on them without breakage, they are likely a good candidate for weaving.

Flax is the source of linen but can also be braided and woven. Bamboo is another that is used for fiber to be spun but can also be used for weaving and basketry. Iris leaves and cattails are good for weaving as well. Even pine needles can be harvested to make baskets.

To harvest and braid or weave leaves, follow a basic formula of drying before use and then wetting as you need to manipulate the fiber. Some will need to be harvested green and dried slowly out of the sun. Others must be harvested when they are dry and no sooner. When the time comes to use the material, either spritz them with water or, if they can handle saturation, lay them in a pan or pot of water for a short time.

Bamboo grows just about anywhere and is edible, hardy, and versatile. Not to mention it adds an exotic flair to any space.
(Photo courtesy of Steve Swayne)

The moistened fiber should be pliable, and you can then manipulate as necessary. Leaves can be braided by folding one in half and laying another in the middle. Slowly and tightly braid the strands until you can secure the starting end against something and really start to tug. Braid tightly without much twist in the strands, and add additional leaves as you run out. A good length for usable rope is 6 feet long.

THORNY MATTERS

Don't wait until the last minute to add new leaves to your braided strand. You'll run the risk of it coming loose. Instead, join a new leaf in the strand when you have a good 6 to 8 inches left.

A finished braid can be woven as desired, or you can simply use dried-then-moistened reeds to weave. Basket making is an ancient art and one that is developed over time. It might start with a table mat and end in full-scale furniture. You probably have something in your garden or backyard now that you could use to get the hang of braiding and weaving. If it appeals to you, plan your garden next year with weavers in mind.

And not every beneficial stalk has to be woven. Strong sunflower stalks can be joined together to make trellises for next year's peas. That is, if you don't just grow the peas to climb the live sunflowers. They can also be grouped together with wire and fashioned to form a fence or shading.

Basket weaving is a rhythmic craft, like spinning, knitting, and weaving. The repetition can be relaxing, all while you are creating something beautiful and useful.
(Photo courtesy of Steve Swayne)

Lip Balm

Soap is not the only hygiene product you can make from your garden. The wax that your bees have been busily building is just right for a lip balm. I wouldn't recommend using it straight out of the hive, though. It's sticky!

After you have harvested the wax, melt 1 tablespoon of the wax with 3 tablespoons of a carrier oil (such as sweet almond, walnut, macadamia nut, sunflower, or avocado oil), plus any butters you'd like (such as cocoa or mango butter). Mix your favorite essential oils into the melted mixture and pour it into containers.

There are lipstick-style molds available from crafting supply stores that you can pour your balm into. Or, an easier option, you can use molds that you have around the house. Consider reusing small containers like metal mint tins, small plastic tubs, or glass vials. (Clean them out before use, of course.)

Once cooled completely (time varies according to the size of the mold), you can store your lip balm at room temperature. Try to resist the urge to eat it. It smells better than it tastes, I think.

Honey isn't the only benefit of raising bees. Their wax can be used in home and health products, freeing you to be self-sufficient in even more ways.
(Photo courtesy of Emma Jane Hogbin)

Candles

Beeswax candles are more than just fragrant. They actually create a static electricity reaction that pulls pollens and other unpleasantries out of the air. Easy and cheap to make (if you source your own wax), this makes candles a must-create with your wax harvest.

The simplest method is to pour melted beeswax into an old glass container that has one or more wicks laying in it, usually with a weight at the bottom to hold them in. A wick, by the way, can be any cord. Hemp works well. Allow the wax to set until completely hardened, and then trim the wicks.

To make tapered candles, cut your wick a little more than twice as long as you need it to be. Dip each end of the wick into the melted wax, and then hang it up by the middle, undipped part of the wick. Make sure it is over something to catch the drips. If you are worried about the wick falling straight, you can tie a washer to each end of the cord to keep it from curling up.

It will only take a minute or so for the new layer to dry. As soon as it dries, dip it quickly again. Too much time in the wax will melt your layers and pull them off. Each dip should make the candle a bit bigger. When they are the size you want, snip the middle of the wick to separate the two and trim the wick down to size so only an inch remains exposed from the wax.

When choosing a container for melting wax, be sure that it is deep enough to dip the candle in, even when you start to run low on wax. Remember that as your candle gets bigger, the pool of wax in your pot will get smaller.

Crayons

We know crayons say "nontoxic," but what is really in those dyes, and where did the wax come from? These are the same crayons your toddler munches on and your preschooler obsessively picks the paper off of. For that matter, what is the point of all that paper if it's just getting peeled away?

If you are harvesting beeswax, you might not be able to answer these questions, but you can ignore them. Beeswax melted with (your homemade) goat's milk soap or glycerin soap makes a perfect crayon. No need to pay for a few rounded, toddler-friendly crayons when you can just make some. Or you can roll them up to familiar shapes and sizes for the big kids.

Grate the soap and melt it while separately melting the beeswax. Carefully mix them together in a 1:1 ratio. Less soap makes a softer crayon; more soap makes a sticky one.

OVER THE GARDEN FENCE

Don't like the color? Melt it down and redye! Miniature slow cookers, as used for fondue, are good for small-scale projects like this. Check your local resale shop for a fondue pot or mini slow cooker on the cheap that you can dedicate to crafting.

When combined, stir in some dye (see "Plant-Based Dyes" later in this chapter if you'd like to try making your own), then pour into tins to set, or let cool and harden a bit so you can roll it into shapes to set. Don't forget to grease anything you pour the crayon mixture into, or pour it onto wax paper.

Homemade crayons are not only safer for your children, but can be made into shapes that fit their age and coordination level.
(Photo courtesy of Steven Depolo)

Flower Cutting

The flowers in your garden serve so many purposes. They attract bees and butterflies to pollinate. They can repel the harmful bugs. Some are edible. All are beautiful. Beauty fades, as we all know, but there are things you can do to make it last just a bit longer. To help cut flowers last longer, keep in mind that they are now only getting nutrients to stay alive from the water you put them in. Maximize that nutrition, and you'll extend their life span. There are a few factors that go into this.

Carefully selected flowers can make stunning arrangements for gifts, crafts, or simply visually freshening up a space.
(Photo courtesy of Steve Swayne)

Flower health. A healthy flower is going to last longer than an unhealthy one no matter how you slice it (pun pretty much intended). So along with being careful cutting the stem and the environment you keep it in, you also need to be selective about the flowers you cut, and then preserve their integrity after doing so. For example, a rose's thorns must be kept intact for the rose to thrive.

Single-stemmed flowers, like daisies, should be fully open when you cut them. If there are multiple flowers on a stem, such as snapdragons, at least one should be bloomed and one closed. If they are all tightly closed, they won't likely bloom in the water. If they are all open, they will not last long.

Morning is best for cutting, before flowers are dehydrated in the afternoon.

Stems. A healthy stem makes a healthy flower. If you hack away at the stem with a dull blade or serrated edge, the vascular system will not be strong enough to support it. Cut at a diagonal angle with a sharp blade. Even after they are in the vase, trim periodically to give them a fresh place to drink. The angled cut helps it stand up in the vase and expose a broad area of the stem to water.

Place your cut flowers in water right away so that they don't dry out or start to wilt.

Environment. Just because they are cut doesn't mean your flowers have quit liking the sun! Be mindful of the sun preferences of your flowers and try to accommodate that.

Flower arrangements brighten your day, and cutting the flowers keeps the plant healthy and growing. It's a win-win all around!
(Photo courtesy of Jen Dickert at SupernaturalNutrition.net)

Also, it doesn't hurt to clean the vase or container before you fill it with water. Bacteria inside the container could keep the flowers from thriving. That also means that changing the water regularly is important. Even though you might like a cold drink in the summer, your plants don't. They soak up the water from soil, which is warm. Florists use lukewarm water, about 100°F.

THORNY MATTERS

Fruits and vegetables give off an ethylene gas that will make your flowers wilt faster. So you might want to rethink your fruit-basket-and-bouquet centerpiece!

Dried Flowers

Have a particular flower or bouquet you want to keep forever? Dried flowers are beautiful in their own right and will usually retain their fragrance.

The simplest method for drying flowers is to tie a string around the stems, whether in bunches or single flowers, and hang them upside down from a hanger or closet rod. Closets are perfect for drying flowers if it is a place where they will be undisturbed. They like sun when they are growing, but when they are drying it is a different story.

After they are dry, spray them with hairspray or acrylic spray to preserve them. Then, they are ready to be displayed. Make a pretty centerpiece by filling a vase with various dried flower petals and foliage. Or leave them on their stems as you would a fresh flower bouquet.

Simple wreaths and centerpieces can be made by purchasing a premade foam wreath or solid shape. The stems poke right into the foam, or you can secure them with a pin. Play with a variety of colors and arrangements based on seasons and circumstances.

This dried wreath was made entirely from homemade dried flowers and would make a pretty home decoration for yourself, or to sell.
(Photo courtesy of Kelby Carr)

To take things a step further, you could fashion a homemade wreath from the stems, stalks, and braided leaves I talked about earlier.

Grapevines make excellent wreaths. Four to 6 feet is good for a standard door wreath, but small ones are great decorative candleholders. Dried flowers and herbs—or fresh ones, depending on the look you are going for—easily and attractively combine with homemade wreaths. These make excellent gifts, or just pleasing reminders of a season well spent on your backyard farm.

Pressed Flowers

Another method of preserving flowers is pressing. Bouquets that have been in a vase for 24 hours are perfect candidates for pressing because they have been hydrated.

Choose the flowers that you will be pressing and arrange them carefully. This likely means laying them flat, without any overlapping. Lay your arrangement on paper. Wax paper, printer paper, and even coffee filters can be good options. Just make sure that you avoid direct contact with anything with a pattern, or the flower will have the pattern pressed into it.

To press in a book, simply put the paper and flower into the pages, close it, and then set something on the book. You can fill an entire book with flowers. Just keep them about an eighth of an inch apart so the roses don't squish into and imprint on the daisies.

Books work for a while, but if you plan to press frequently, consider buying or making a dedicated flower press.
(Photo courtesy of Brian Boucheron)

It will take several weeks for the flowers to dry fully, so if you want to speed it up a little, consider a microwave press kit. There are relatively inexpensive models available, or you can make one from a cardboard–newsprint–flower–newsprint–cardboard sandwich, secured with rubber bands. Microwave briefly on a medium setting, and then remove them to cool with a book or something weighing them down. Repeat heating and cooling until they are completely dry, or you can simply leave them under the book for a couple of weeks. Not lightning fast, but an improvement nonetheless. Always use caution with this method, and never leave them in the microwave unattended.

When your flowers are pressed, try dabbing a bit of glue on the back and sticking them to cardstock. Pressed flowers make lovely stationery and greeting cards. You could also decoupage them to frames and boxes, or layer them between contact paper to make placemats or bookmarks.

Pressed flower crafts are incredibly kid-friendly and yet another way you can bring the whole family together over your garden harvest.

Lavender Wands

Years ago, lavender wands were used to freshen up fabric. They were tucked inside stored linen to keep it smelling fresh, and it might have even helped with bugs. Today, lavender wands are just as fragrant and useful—and they might even double as a fairy wand for a little one or an addition to a flower arrangement for a special someone.

Weave lavender stalks into wands.

With an uneven number of lavender stems, leave the flowers at the top intact and clean off any little leaves or buds along the bottom half of the stem. Work quickly, before they become dry and brittle.

Hold them with the flower heads even with each other, and tie a long, narrow ribbon around them at the base of the flowers. I would leave the length of the ribbon extra long so that you don't wind up running out. The tail of the knot can tuck right into the bunch of stems. Tie this nice and tight so that it doesn't come loose or undone.

These lavenders have round, thick flower heads while other lavender varieties have long, thin flower spikes. Both will work for lavender wand weaving.

And this is why you need them to bend but not break: taking two stems at a time, fold them down over the knot and wrap the long end of your ribbon around them. Bend two more stems down, but this time on top of the ribbon. Again, fold two down and wrap the ribbon on top of them. Keep folding the stems and weaving the ribbon this way—over, under, over, under—until you have folded them all down. The last fold will be three rather than two, but it will mark your place and work itself out in a minute.

On the next row, alternate them. If you went over on a pair, go under. If you went under, go over. This will create a pretty pattern as well as secure them tightly.

Circle around for a few rows, and then switch to going over and under single stems rather than pairs. Keep the ribbon as close to the row above as possible, and the flowers inside. By pulling the ribbon progressively looser toward the middle and tighter as you get to the end of the stems, you will create an oval shape.

Fold the stems down as you weave the first row.

Weave every two or three stems for the first part, then weave every other stem where the middle is larger over the thickest part of the flower heads.

Toward the end, switch back to weaving over and under pairs instead of single stems. When the flowers are gone and you are back to stems only, wrap the ribbon around a few times tightly and then tie it off. The tail can now be cut and the knot pushed inside the stem. If you'd like, you can tie an extra bow with long tails, or stems wrapped down the length of the wand. Just make sure it's secured so that the sprigs don't pop loose.

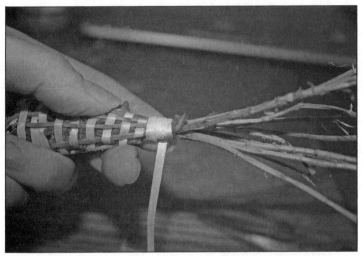

Secure the end with tight wrapping and a knot. You could also secure with a drop of hot glue or crafting glue if you want. We usually don't.

And there you have it! Your lavender wand will dry this way, and you will have fresh-smelling lavender that will last for years.

Plant-Based Dyes

The most dramatic effect your flowers can have is to be used for dye. It's probably also one of the oldest uses. Fabric has been dyed for centuries, long before packets of powder were sold at the supermarket. Even today, commercial dyes find their beginnings in plant and animal matter.

And it's not just your flowers. Roots and berries and leaves (and pretty much anything) can be used to make dye. Now, some plants have already been proven over time to give certain colors and intensities. But that doesn't and shouldn't stop you from exploring and trying your hand at new combinations.

Dye making is very simple. Just chop up the material you would like to turn into dye and put it in a pot of water. Twice as much water as material is a good ratio, but you can add a little more plant matter for a more saturated color, or a little more water for a lighter color. Bring to a boil, then simmer for an hour. That's it. The water you strain out of the pot will be your dye bath, and you can submerge anything in it you'd like to color. Or, you can use it in your crayon and soap making.

It will help to premordant the fabric—that is, treat the fibers with a solution that will prepare it to hold color longer and more vibrantly. Vinegar is good for most dyes, though salt works well for berries.

Butternut seed husks, juniper berries, dandelion roots, goldenrod, and carrot roots are just some of the plants that will yield shades of orange and brown. Berries and certain iris blossoms can create purples and pinks, and you should be able to get a lovely red out of roses and hibiscus. Roots and walnuts will make black. And if you have ever boiled or steamed artichokes you know that they create green, as does nettle and chamomile. (You may be thinking of using beets for their red color, but they won't bond to fibers well and can stain the skin.)

Of course, plant-based dyes are not guaranteed to look the same every time. A lot depends on the quality and quantity of the plant, what you are dyeing, and any additives that you combine with it. There will be variants, but that is what makes it beautiful.

Simple Plans
for the
Backyard Farm

Here are three simple plans for your backyard farm. The cold frame will be useful for year-round gardening, the chicken brooder can be used to hatch your own chicks or brood mail-order chicks, and the goat pen is a versatile open shelter with myriad adaption possibilities.

Cold Frame

Start seeds outdoors earlier in the spring, and keep your plants growing longer in the fall with a simple cold frame. This plan uses recycled windows from a home remodel to cut material costs. Recycling throwaways like this will save you money in your gardening efforts. Our window was 35½×27½", but your window might be different. Older homes are notorious for having odd-sized windows.

Supplies:

> 1 wood-framed window
>
> 1 8' 1×10 board
>
> 1 8' 1×12 board
>
> 2 medium hinges
>
> Small box of 1" screws or nails
>
> Screen door or small handle
>
> 1 2' 2×4 board

Steps:

1. Cut two lengths of 1×10 28" long and one length 34" long. Cut one length of 1×12 to 35½" long (this is the front).

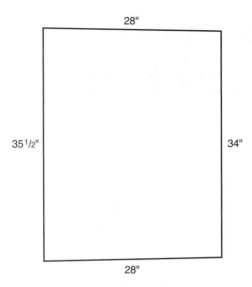

28"

35 1/2"   34"

28"

Front view of a cold frame.

2. Screw the side pieces (28"-long boards) to the back board (34" board) making sure that the back board is mounted on the inside of the side boards (see following diagram). Use at least three screws at top, middle, and bottom of board. You can drill a pilot hole one size smaller than the screw to prevent spitting the board.

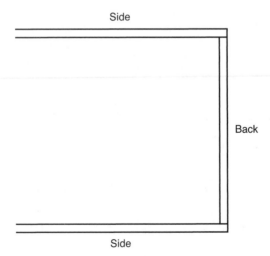

Side

Back

Side

Screw side pieces to the back board.

3. Attach the front board to the sides. The side boards should be inside the front board. Again, use pilot holes to prevent splitting the boards.

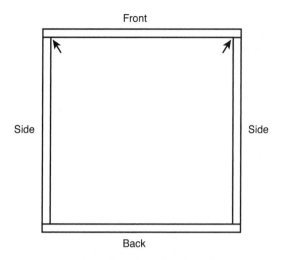

Attach the front board to the sides.

4. Cut four pieces of 2×4 4" each and place them on the inside corners of the box. Attach them so they are flush with the top of the box and screw them into place (these help stabilize the box).

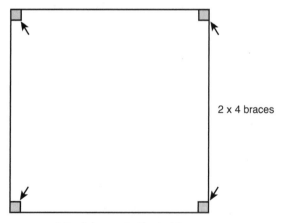

Attach braces to the inside corners.

5. Place the window on top of the box. Attach hinges to the window and back of the box. Your hinges probably came with screws for the hinges so use those screws. Attach the handle to the top of the window's wooden frame at the front so you have an easy way to open the cold frame when needed.

Side view (from back hinges)

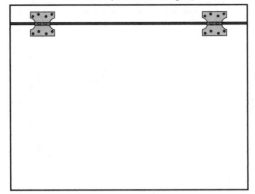

Attach hinges.

Chicken Brood Coop

A simple chicken coop with a homemade door and slanted roof provides three sides of shelter. We used a sheet of tin we salvaged from an old barn for the roof of this brood coop. It can be used to house a setting hen while she incubates the eggs, as a transitional pen for baby chicks when you bring them outdoors, or as a quarantine coop for new arrivals to your backyard farm.

Supplies:

> 1 4' sheet of sheet metal
>
> 8 2×4s, 8' long each
>
> 1 4×8 sheet of ½" or ¾" USB/plywood board
>
> 2 small hinges
>
> 1 door latch
>
> 1 box of 1" screws

Steps:

1. Cut plywood 3' wide and 44" long for the floor of the coop. Cut two 2×4 posts at 36" and two posts at 24" to serve as the legs. Cut two 2×4 posts 44" long for braces at the bottom of the cage and two more 48" long for the top. Cut eight braces from 2×4s that are 36" long for bracing the floor and roof.

2. Build the right side of the coop first. Attach the 48" top brace on the top of the two legs, one 24" in the front, and one 36" in the back. (This is easier to do if you lay it down on the ground.) Attach the 44" post on the inside of the legs (see the following diagram). Repeat on the left side so you have two sides built.

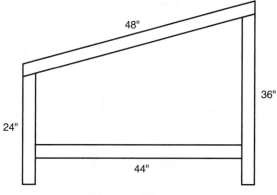

Side view of the coop.

3. Add the floor braces on the inside of your chicken coop to secure the two sides together. Start with the front and back braces and then work in the middle two braces. Braces will fit inside the bottom brace to prevent gaps in the floor where mice or snakes could enter. Put the floor in place and secure before adding top braces. Attach the top braces.

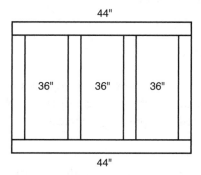

Bottom view of the coop.

4. Add braces for a door in the center of one side. Measure your door according to your needs (remember, one brace will be shorter than the other because of the sloped roof). Our door was about 12" and barely fits our small waterer; 12–18" is a good size opening. Cut your

plywood door to match the opening you've created and attach with hinges on one side and a latch on the other.

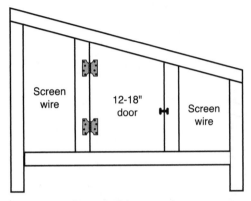

Side view of the coop door.

5. Attach your screen and plywood walls as desired. A broody hen wants to feel protected and private, so we built three plywood walls and only kept a screened wall on the side with the door.

6. After all braces and screening are attached, screw or nail the 4' piece of sheet metal to the top. Be sure to leave a 2" hangover on both ends of cage to allow water to shed properly. Enclose back of cage with plywood to protect from weather. We use a tarp during heavy spring storms to cover the screened wall.

Goat or Sheep Pen

A four-sided shelter with an open doorway, this goat or sheep pen provides a warm, dry area for your livestock but is simple enough to build in a single afternoon. Note: You can use plywood instead of tin, but tin will not need waterproofing treatment.

Supplies:

5 sheets of tin per side (will need to cut to fit without hanging over)

5 sheets of tin for top

9 posts cut at 10' long

15 10' 2×4s

1 box of 16 penny nails or screws of same length (your preference)

8 10' 1×4s

Steps:

1. Set corner posts 2' deep in front of the shed. Trim 2' off back posts and set those 2' deep as well. This will make your back two corner posts 6' tall and your front two posts 8' tall.

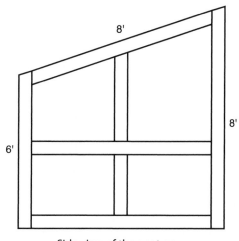

Side view of the goat pen.

2. On the front of the shed set middle post for door frame 7' from one of the corners to give yourself a 3' open doorway.

3. On the back and sides of the shed set your middle posts in the direct center between the two corner posts.

4. Add the 7' braces to the front of the shed at the top, middle, and bottom of the shed wall area. Be sure they are level.

5. Add your 8' braces on the side of the shed at the bottom and middle. You'll need to measure the length for the top brace because the sloped roof will make it slightly longer than 8'. Repeat on the other side.

6. Add the 10' braces at the top, middle, and bottom of the back of the shed area.

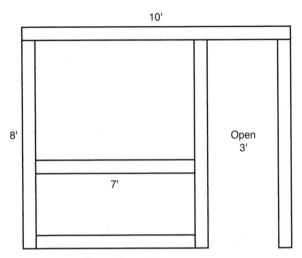

Front view of the goat pen.

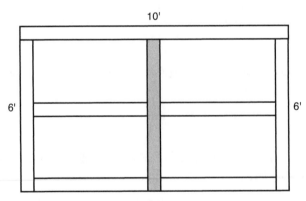

Back view of the goat pen.

7. You can use plywood to create the shed walls, securing the sheets of plywood to the braces and corner posts with screws or nails. Tin sheeting can also be used; it often comes in 3' wide sections by 8' long. Trimming the wall tin lengths with tin snips will be necessary. Overlap the ridges of one sheet of tin with the ridges from the previously attached sheet of tin to avoid gaps in your wall or roof.

8. Add the rafters on your shed. Space them no farther than 2' apart. You can use five 10' lengths of 2×4s turned on their sides to create the appropriate length for your roof to shed water. Add 1×4 boards across the rafters (they will run from side to side) for the roof to attach to. Space those 1' apart from front to back. This is what the roof will attach to.

9. Whether using tin or plywood, leave the overhang at the front and back of the shed roof to provide clearance for water runoff. This is why we used 10' boards for the rafters.

Gardening Journal Pages B

These garden journal pages may be copied for your home use, or you can find downloadable versions online at BackyardFarmingGuide.com.

The **Garden Layout** provides a simple place for you to sketch the garden layout of your vegetable garden, herb garden, or front border areas. Be sure to number or name each garden section so you can track which plantings go where.

The **Seed Starter Log** helps you track which plants and varieties you tried this year, where you purchased your seed, and how the plants performed for you. I use this page to note seedlings and transplants that I purchase as well.

The **Monthly Garden Journal** gives you a place to jot down chores to do and note anything that happened this year. Were there any pests that were particularly bad? Did you have an empty space in your garden between spring and summer plantings that you'll want to be sure to fill next year? Each month in the garden brings something new and exciting, so print or copy one sheet for each month of the year.

The **Notes for Next Year** is for tracking the things you'll want to do differently next year. Whoops! You should have planted those seeds a lot sooner. Next year try a drip irrigation system and see if that cuts back on your watering bill. All the things you won't otherwise remember in January of next year go here!

Garden Layout

www.BackyardFarmingGuide.com

Garden Bed # ....

Seed Starter Log

www.BackyardFarmingGuide.com

Seeds	Planted	Germinated	Transplanted	Notes

Monthly Garden Journal

Plants Blooming or Harvested

Pests & Beneficials Observed

Action Taken

Garden Observations

Notes for Next Year

www.BackyardFarmingGuide.com

Resources   C

One thing all backyard farmers know, whether operating on a large or small scale, is that information is power. In this appendix, I've recommended several of my favorite books and websites, organized by section, so you can find more detailed information on some of the concepts in this book. A more extensive listing of resources and full reviews of materials can also be found at BackyardFarmingGuide.com.

Part 1: Living Large on a Small Scale

Food, Inc. takepart.com/foodinc.   This movie is chock-full of statistics, quotes, and behind-the-scenes glimpses of what the current, industrialized food production looks like in America today. Once you've watched this movie (I saw it through Netflix) you will understand the importance of eating locally and knowing where your food is being produced.

Seymour, John. *The Self-Sufficient Life and How to Live It: The Complete Back-to-Basics Guides.* New York, NY: DK Publishing, 2009.   A good general overview of homestead living, I love Seymour's philosophy of self-sufficiency in the first chapters.

Ussery, Harvey. *The Small-Scale Poultry Flock.* White River Junction, VT: Chelsea Green Publishing, 2011.   The first section of this book shares the current system that is so common and how this system is sick and unsustainable.

Part 2: Gardening on a Backyard Farm

Ashworth, Suzanne. *Seed to Seed: Seed Saving and Growing Techniques for Vegetable Gardeners.* Decorah, IA: Seed Savers Exchange, 2002.   The most complete manual on seed saving that I've seen yet with details about distances needed between varieties, and how to save seeds from more obscure plants.

Bartholomew, Mel. *Square Foot Gardening*. Brentwood, TN: Cool Springs Press, 2006.   A classic reference for those who are interested in the square-foot style of gardening, I still use this book as a spacing reference for my intensively planted rows.

Heffner, Sarah Wolfgang. *Heirloom Country Gardens*. Emmaus, PA: Rodale Press, 2000.   A beautiful book featuring old-fashioned vegetables, herbs, and flowers.

Markham, Brett. *MiniFarming: Self-Sufficiency on ¼ Acre*. New York, NY: Skyhorse Publishing, 2010.   Heavily focused on building soil fertility and sustainable, intensive gardening techniques with less information on the plants themselves.

McLaughlin, Chris. *The Complete Idiot's Guide to Heirloom Vegetables*. Indianapolis, IN: Alpha Books, 2010.   Growing and seed saving information as well as descriptions of several hundred heirloom vegetable varieties.

————. *The Complete Idiot's Guide to Small-Space Gardening*. Indianapolis, IN: Alpha Books, 2012.   Great information for squeezing the most produce out of every space possible. The smaller your garden area, the more you need this book.

Phillips, Michael. *The Holistic Orchard: Tree Fruits and Berries the Biological Way*. White River Junction, VT: Chelsea Green Publishing, 2012.   One of the smartest books about keeping fruit organically that I've ever read, this is a must-have for backyard farmers who want productive, organic fruit trees on their property.

Rodale, Maria, Anna Kruger, and Pauline Pears. *Rodale's Illustrated Encyclopedia of Organic Gardening*. New York, NY: DK Publishing, 2005.   Overview of organic gardening techniques, pest control, and various vegetables, fruits, and herbs. Doesn't limit itself to heirloom varieties.

Solomon, Steve. *Gardening When It Counts: Growing Food in Hard Times*. Gabriola Island, BC, Canada: New Society Publishers, 2005.   This book also largely focuses on increasing and maintaining the soil fertility and includes some well-researched information about plants I haven't seen elsewhere. Contains the formula for "Complete Organic Fertilizer" mix to feed your plants in a pinch.

Part 3: Animals for a Backyard Farm

The American Livestock Breeds Conservancy. albc-usa.org/heritagechicken/index.html. Threatened or endangered heritage chicken breeds and their best roles in the backyard farm.

Belanger, Jerry. *Storey's Guide to Raising Dairy Goats*. North Adams, MA: Storey Publishing, 2010.   More specific information about the keeping of dairy goats, especially when it comes to freshening the goats on a yearly basis and dietary concerns of a heavily producing dairy goat.

Bonney, Richard E. *Beekeeping: A Practical Guide*. North Adams, MA: Storey Publishing, 1993. A great basic book that includes good information about the honey harvest and potential ailments bees face.

Deleplane, Keith. *First Lessons in Beekeeping*. Hamilton, IL: Dadant & Sons, 2007. A seemingly simple little book, this offering by Dadant's apiary supplies has everything you need to get started.

Ekarius, Carol, and Paula Simmons. *Storey's Guide to Raising Sheep*. North Adams, MA: Storey Publishing, 2009. A good overview of sheep ownership with specifics of wool production, livestock guardians, and how to obtain an organic certification.

Practical Rabbit Housing Brochure by Southern University Agricultural Center. bit.ly/BYFRabbitHouse. Compares several hutch styles.

Sayer, Maggie. *Storey's Guide to Raising Meat Goats*. North Adams, MA: Storey Publishing, 2010. Information about marketing meat goats and specializing in making a profit from meat goats.

Part 4: Enjoying the Bounty

Barry, James, and Margaret Floyd. *The Naked Foods Cookbook*. Oakland, CA: New Harbinger, 2012. Whole foods recipes that incorporate many of the freshly grown or locally available produce you want to eat.

Bubel, Mike, and Nancy Bubel. *Root Cellaring: Natural Cold Storage of Fruits & Vegetables*. North Adams, MA: Storey Publishing, 1991. An excellent resource for storing foods in a root cellar or basement storage.

Gasteiger, Daniel. *Yes, You Can! And Freeze and Dry It, Too*. Brentwood, TN: Cool Springs Press, 2010. A great resource that is highly visual and shows clear step-by-step directions for canning, freezing, dehydrating, and pickling your garden's produce for longer-term storage.

Gladstar, Rosemary. *Herbal Recipes for Vibrant Health*. North Adams, MA: Storey Publishing, 2008. If you only get one herbal medicine book, this is one that is worth your money to buy. Dozens of recipes as well as an explanation of how to prepare herbal remedies in the appendix section makes this book a good reference.

Part 5: Crafting from the Backyard Farm

Berry, C. J. J. *First Steps in Winemaking*. East Petersburg, PA: Fox Chapel Publishing, 2011. Lots of information about making your own wine, including grapes for your area and favorite wines, producing the juice, fermentation, and the intricacies of bottling and aging wines.

Callahan, Gail. *Hand-Dyeing Yarn and Fleece*. North Adams, MA: Storey Publishing, 2010. This book covers the coloring process of the yarn or fleece but not the spinning information. Together with the following book you'd get a complete look at the total process.

Casey, Maggie. *Start Spinning: Everything You Need to Know to Make Great Yarn*. Loveland, CO: Interweave Press, 2008. From working the fleece, blending fibers, and spinning various yarns, to selling your wares, this book covers the stages of fleece to yarn in-depth.

Emery, Carla. *Encyclopedia of Country Living, 10th Edition*. Seattle, WA: Sasquatch Books, 2008. Covers a wide variety of country living tips and tricks. It's like reading 20 years' worth of garden journals by a masterful homesteader.

Gehring, Abigail R. *The Illustrated Encyclopedia of Country Living*. New York, NY: Skyhorse Publishing, 2011. Huge resource book for the lost arts of homesteading such as cheese and butter making, soap making, and floral arts.

Pooley, Michael, and John Lomax. *Real Cidermaking on a Small Scale*. East Petersburg, PA: Fox Chapel Publishing, 2011. Details about making cider at home, including specific guides to apple variety selection for different flavors.

Index

dyeing fibers, 339
  acid dyes, 339
  basic recipe, 339
  caution to washing, 340
  depigmentation, 339
  dye type, 339
  fleece, premordant, 340
  hand painting with dye, 341
  hands-off dyeing, 340
  natural dyes, 339
  preventing felting, 339
  roving texture, 339
  slow cooker, 340
  stovetop or vat dyeing, 340
dyes, 377-378

E

ear mites, rabbits, 211
Early Prolific Straightneck squash, 123
Early Purple Sprouting broccoli, 103
Earth-friendly lifestyle, 5
earworms, 114
East Fresian sheep, 215
echinacea, 131, 317
Echinacea purpurea, 131
edible landscaping, 55
  climbing vines, 55
  flowering perennials, 55
  front yard, 32
  ground cover, 55
  landscaping shrubs, 55
  lavender, 132
  ornamental annuals, 55
  shade trees, 56
eggplant, 54
eggs, 5, 197
  aging flock, 198
  appearance, 181
  beginner eggs, 197
  collection, 198

color, 198
dual-purpose breeds, 197
misshapen, 197
molting, 197
myth, 197-198
nutrition, 181
pullet eggs, 197
which hens are laying, 198
elderberries, 164-165
Elfin thyme, 144
English Angora rabbit, 206
English Morello cherry, 149
English Munstead lavender, 55
equipment, cost savings, 12
Erica's Inside-Out Lasagna, 289
ethylene gas, 302
European corn borer, 97
European grapes, 158
ewe, definition, 216

F

fabric. *See* fibers
Fair Maid of Devon apple, 346
Fairy Tale eggplant, 54
Fantasia nectarine, 152
Farmer Brown Shears His Sheep: A Yarn About Wool, 333
farmers' markets, 27
felting, 337
fencing, 21, 64
  fence pliers, 66
  materials, 38
  posthole digger, 64
  undeveloped land, 16
  vinyl, 22
  wire cutters, 66
  wood panel, 21
fermentation
  cheese making, 328
  cider, 349

heirloom plants, 173
home brewing, 345
raw cow's milk, 322
wine, 352
yogurt, 325
fertilizers
  high-nitrogen, 124
  organic, 79
  petroleum-based, 6
  synthetic, 103
  tomato, 124
fibers, 333
  dyeing, 339
    acid dyes, 339
    basic recipe, 339
    caution to washing, 340
    depigmentation, 339
    dye type, 339
    fleece, premordant, 340
    hand painting with dye, 341
    hands-off dyeing, 340
    natural dyes, 339
    preventing felting, 339
    roving texture, 339
    slow cooker, 340
    stovetop or vat dyeing, 340
  felting, 337
  fiber arts, 338
  fleece, 334-338
    animals, shedding season, 336
    carding, 337-338
    coarse hairs, 336
    combing, 338
    crimp, 335
    drum carder, 338
    factors, 334
    fleece characteristics, 334
    lanolin, 337
    longer fibers, 335
    shade, 335
    skirting, 336
    washing, 337